UNDER THE HIGH SEAS

UNDER THE HIGH SEAS

New Frontiers in Oceanography

by Margaret Poynter
and
Donald Collins

ILLUSTRATED WITH PHOTOGRAPHS, CHARTS AND MAPS

ATHENEUM · 1983 · NEW YORK

Library of Congress Cataloging in Publication Data

Poynter, Margaret.
Under the high seas: new
frontiers of oceanography

SUMMARY: Discusses folklore, exploration, and composition of oceans; their use for transportation and as a source of food, oil, minerals, and energy;
and international law concerning oceans.
1. Oceanography—Juvenile literature. [1. Oceanography] I. Collins, Donald, joint author. II. Title.
GC21.5.P69 1983 551.46 82-16338
ISBN 0-689-30977-5

Published simultaneously in Canada by
McClelland & Stewart, Ltd.
Composition by Service Typesetters, Austin, Texas
Printed and bound by Fairfield Graphics,
Fairfield, Pennsylvania
Designed by Mary Ahern
First Edition

To BARRY,
who crossed an ocean
to find us

Contents

1

"Here Be Dragons"

EARLY ON a spring day in the year 1804, a Spanish frigate was sailing just north of the Azores. The ship's lookout was scanning the horizon when the outline of another craft appeared.

"Brigantine off the port bow!" he shouted. "She's proceeding northwest." He peered through his glass. "No sign of a crew on deck. Perhaps she's in trouble."

The frigate's captain ordered his crew to haul wind. "Prepare to lower a boat," he said. "We're going to board her."

Thirty minutes later, the frigate's first mate and two other sailors were clambering onto the deck of the brigantine. The name *Maria Therese* was inscribed on the bow in bold black lettering. The ship's one lifeboat was gone, and all the hatches were uncovered. The hold was full of barrels, and all of the ship's sails were set, with the exception of one, which was torn and flapping in the wind. An inspection of the masts and spars showed them to be sound. The wheel was not fastened, but in good working condition.

The three men went belowdecks and found that the chronometer, sextant, and navigation and log books were in place in the captain's cabin. The ship's stores contained enough food to last for six months, and there was plenty of drinking water. When the sleeping quarters were searched, the men found chests full of clothing. Boots, oilskins, pipes, and tobacco had all been left behind by the brigantine's crew.

Under a bunk the first mate found a sword that had a faint brownish stain on its blade. His skin crawled with apprehension as the dank, musty odor of the vessel filled his nostrils. What's happened here? he wondered. Mutiny? Murder, perhaps? Is this craft inhabited by evil spirits? Why would a crew desert a ship that's fit to circle the globe?

"I've seen enough," he said to his crew mates. "Let's get out of here."

The two sailors were more than willing to obey his command. They were certain that the *Maria Therese* was a ghost ship.

Their relief was short-lived, because their captain decided to bring the derelict into port. "She'll bring a handsome salvage fee," he said. "And there'll be an extra mug of grog for each sailor that mans her."

The bonus didn't allay the fears of the seven men who were selected to board the *Maria Therese*. As night fell, their feelings of dread grew. They sailed through the darkness, wondering if they would still be alive when the sun rose over the horizon.

Meanwhile, aboard the frigate, the lookout kept a close eye on the brigantine until just before dawn. At that time a dense fog blanketed the region. With the rising of the sun, the fog disappeared, and the frigate's captain was relieved to see that the *Maria Therese* was still close by. He grew troubled, however, when he could see no one on deck. He shouted through his megaphone, but received no answer. He shouted again. Still no response. Fear gripped his belly.

"Something's amiss," he said to his first mate. "Prepare to board her again. This time I'll accompany you myself."

A brief search of the derelict revealed that the seven men had vanished. The only signs that they had been there were some dirty pots and plates in the galley, and some half-empty mugs on the dining table.

The captain quickly decided to forego the salvage fee. He wanted nothing more than to leave this evil ship behind. He wanted to forget that he'd ever seen her. In his heart, however, he knew that he would always be haunted by the memory of the *Maria Therese*. After all, because of that accursed ship, hadn't he sent seven good and faithful men into oblivion?

THE LEGEND of the *Maria Therese* had its beginning on the high seas. Just a little over fifty years earlier another such legend was born off the shore of New England. At that time, on a calm sunny afternoon, the sloop *Seabird* dropped anchor near the entrance to Narragansett Bay. The captain of a passing fishing vessel exchanged signals with the captain of the *Seabird* and was assured that all was well aboard the sloop.

The following morning all was definitely *not* well. The *Seabird,* under full sail and seemingly bent on self-destruction, was speeding straight toward the coast. Its suicidal journey ended when it ran aground with a crash of splintering timbers.

Several men rushed from the nearby village to offer aid to the crew. They were aghast when they found that the only living creature onboard the hulk was a dog. The *Seabird*'s captain and crew had vanished, never to be seen or heard from again.

WHAT HAPPENED to the crew of the *Seabird* and of the *Maria Therese*? What happened to the vessels that disappeared with all hands, leaving behind them no wreckage, no sign that they ever existed? Are there creatures in the blackness of the ocean's depths that lie in wait for such prey?

The men who sailed the sea in the early days of ocean travel wanted answers to these questions. In most cases, they could find none. If a crew had to abandon ship, or if a vessel were attacked by pirates, there was no radio over which a message could be sent. If a craft was destroyed by wind and wave, months could pass before it was even reported missing. Meanwhile its timbers and cargo would be scattered by currents, and its crew would disappear without leaving a trace.

No man had ever seen what lay more than a few feet beneath the surface of the sea. As far as anyone knew, the darkness could be hiding the most unimaginable kinds of monsters. There really could be a Davy Jones' locker, where drowned sailors were punished for their wrongdoings.

Since they didn't have any answers, and since voyages often took years to complete, sailors spent their time making up their own explanations to these sea mysteries. The "yarns" might start out small, but with each retelling more imaginative details were added until they became full-blown legends. There were tales of monstrous whirlpools that could swallow entire ships; of giant squid with arms long enough to encircle the largest of vessels and drag it to the seafloor; of evil spirits that could cause a crew to vanish; of an angry God that could punish a man by making him sail on forever with a crew of dead men. Sailors often reported that they had seen such a phantom ship on the horizon.

The map makers of the day did their part in spreading the fear and superstition. On their charts, when they outlined an unexplored region, they labeled it with the ominous words, "Here Be Dragons."

A dread of the unknown lay in the heart of every sailor, but that fear didn't stop men from going to sea. If they wanted to conduct trade, to collect the bounty of the ocean, or to wage warfare, they often had little choice. As they traveled farther and more often, they realized that some of their fears had sprung from ignorance. The more they learned about navigation, about the effects of currents, winds and tides, and about the locations of dangerous reefs and shoals, the more control they had over their fate.

Today's sailors have radar, sonar, instant communication with land and with other ships, advance storm warnings, accurate maps, current and tide charts, and modern navigational techniques. Ocean travel has become much less of a game of chance. When a ship leaves port, its crew can be confident that it will arrive safely at its destination. It would seem that man has mastered the ocean.

That mastery, however, hangs on a very thin thread. Human error, carelessness, and just plain bad luck can still cause a ship to sink. A severe storm can tear apart even the largest, most seaworthy vessel. In February of 1982, a 12,500-ton tanker broke up when it was caught in heavy seas. Fourteen men were lost, and the rest of the crew had to cling to the stern until helicopters could rescue them.

Sailing vessels are still at the mercy of wind and wave. Recently, a ketch carrying seven men left Providence, Rhode Island, bound for a site about 600 miles offshore. The craft's navigational radio stopped functioning after only one day, and when a storm rose a day and a half later, the emergency engine didn't work. The wind ripped the mainsail, and there was no spare sail and no repair kit.

During a period of relatively calm weather, the crew used ordinary sewing thread to mend the canvas as well as possible. A short while later, the ketch was being tossed about by hurricane force winds. When the storm abated, the fifty-two-foot craft had been driven off course and was caught up in the powerful current of the Gulf Stream. For three days, the men found themselves being carried further and further out to sea.

The cruise was supposed to have lasted only a week, but seventeen days passed before the men sighted Providence again. As they neared shore, the continuing force of the storm drove their craft past the harbor entrance. After narrowly missing the jetty, they were forced to anchor in a small nearby bay with no docking facilities, called a harbor of refuge. They then went ashore in the ship's lifeboat.

That vessel got into trouble because of a combination of mechanical failures and human error. Other ships seem to get into precarious situations for no apparent reason, except that they were at the wrong place at the wrong time.

Some parts of the sea have more than their share of these unexplained occurrences. One of those infamous regions is the Bermuda Triangle. It's bordered by Florida, the islands of the West Indies, and Bermuda, but many accidents occur far beyond those boundaries.

The mystery and fear surrounding the Bermuda Triangle began during the earliest days of exploration. Columbus reported seeing strange, glowing white water, and an unusual-looking whirlwind near Haiti. Since then, there have been hundreds of reports from people who saw balls of fire, were chased by clouds, and threatened by incredible magnetic storms. Many other survivors of the Triangle have also said that their compasses started spinning, their instruments stopped working, their electrical power was drained, and they lost their sense of time and direction.

Hundreds of people have entered the Bermuda Triangle and never reappeared. On December 5, 1945, five Navy patrol planes left Florida on a routine mission. The weather was clear and the sea calm. The flight was almost over when a frantic message came to the control tower from the lead pilot.

"We seem to be off course," he said. "We cannot see land."

"What is your position?" the controller radioed back.

"We can't be sure just where we are. We seem to be lost," came the confused reply.

The tower operators were startled. How could five

planes in excellent condition with experienced crews be lost in familiar territory in good weather conditions?

"Assume bearing west," the controller ordered.

"We don't know which way is west," said the pilot. "Everything is wrong . . . strange. We can't be sure of any direction. Even the ocean doesn't look as it should." He was quiet for a moment. "It looks as if we're entering white water. We're completely lost." With these words, he stopped transmitting.

By that time a skilled thirteen-man rescue team was on its way to search for the missing planes. The team's plane also disappeared. The search for the six missing aircraft continued for five days, but not one piece of wreckage was found that could furnish a clue to their fate.

"They vanished as completely as if they'd flown to Mars," said a Naval officer. "We don't know what is going on out there."

WHAT *is* GOING on out there? Some logical explanations have been given for the strange sights, the eerie sense of dislocation, and the apparent time warps that occur in the Bermuda Triangle. The area is well known for its treacherous, merging currents, its violent and unexpected changes in wind direction, and for sudden storms in which water and sky join together to form a continuous, confusing "white-out." The "whirlpool" that Columbus reported could have resulted from a waterspout or a hurricane formation. His "glowing white water" was probably caused by the bioluminescence, a glow made by some small marine organisms and used as a defense mechanism.

There are many known cases of piracy and highjacking in the area, so some of the ship disappearances could have been the result of criminal activity. Just the fact that there is such a great volume of sea and air traffic in and near the Triangle would account for more than the usual number of accidents.

Nevertheless, even after considering all of the obvious and not-so-obvious explanations, scientists, experienced pilots, and veteran sailors agree that odd things do happen in the Bermuda Triangle and in other similar areas of the globe. The lack of understanding probably comes from the fact that in many ways the world ocean is still largely a mystery to us. Only a small part of the total sea floor has been accurately mapped. We don't understand the sudden shifts in some ocean currents. There are no doubt some marine creatures that man has never seen.

The truth is that we know a lot more about some parts of outer space than we do about the seas of our own planet.

2

From Skin Canoe to Supertanker

Our ignorance of the world ocean leaves a big gap in our understanding of our environment, because over two-thirds of our planet is covered with water. Until recently, most of our efforts to learn more about the sea were centered about our need to get from one shore to another. To explore, to fish, to conduct trade, to conquer new territories, man had to find ways to travel over the water's surface more quickly and easily. To do so, he not only had to learn about currents and winds, but also to build better ships.

The earliest water craft was only a log that was paddled by hand. It was soon found that binding several logs together with vines or crude rope provided a much more stable craft. Later, men scooped out the insides of tree trunks to make canoes and used paddles shaped from wood to power them. Still other people made vessels from reeds or wooden frames covered with animal skins or bark.

Perhaps no primitive craft was as suited to the needs of its designer as the kayak. As a result, it is still being used today. Slipping into its snug hatch, the Eskimo laces the skin deck tightly about his waist. As he paddles, the man becomes an extension of his boat. Using the forward or backward motion of his body, he can move in one direction or another. When overturned, he uses the paddle to right himself.

Amid the wind-driven ice floes of the Arctic, the light, fragile structure of a kayak is much more serviceable than that of a sturdy wooden boat. An impact that would shatter solid planking nudges a kayak aside and leaves it undamaged. To travel among the rifts in the ice pack, the hunter must often portage from one stretch of open water to another. When he does, he can easily carry his kayak in one hand until he comes to another rift in the floe and is on his way again.

ONCE MAN had mastered the familiar inland waterways, he longed to explore, to conquer distant lands, and to increase his trade with other people. As early as 1300 B.C., the Phoenicians were experimenting with new ship designs. By using a square sail, with which they could run before the wind, they were able to control the trade routes of the Eastern Mediterranean for many centuries. Their ships proved to be far superior to the slave-powered galleys that were developed by the Romans. Those beautiful ships had graceful lines and elaborate decorations, but they were slow and their interiors were full of human misery. Men spent their lives chained to their posts, half starved and covered with vermin. If they lagged behind, they were whipped. When they died, they were simply tossed overboard.

The Vikings later adopted the Phoenicians' square sail and added a keel, a ridge below the center of the boat that extends from stem to stern. The keel added stability to the vessel and with its help the Vikings were able to journey all the way to the shores of North America. Their trip was an amazing feat in the days of "dead reckoning," a crude navigational technique in which a ship's progress was estimated by its presumed direction and rate of speed. On an unknown course and with no land in sight, dead reckoning was a risky method of getting from one place to another.

In the early 1400s, the Portuguese developed the art of celestial navigation, in which the stars were used to set a course. Prince Henry the Navigator established an academy of seamanship and encouraged voyages of discovery. As a result, Portugal became the leader in sea travel. This leadership was lost when Columbus, sailing under the Spanish flag and using celestial navigation, tried to find a faster way to get to India. The fact that he actually ended up on the shores of North America didn't detract from the success of his voyage.

England soon became Spain's biggest rival. These two countries, with their advanced navies and merchant fleets, managed to divide most of the spoils of the New World between them.

The men who sailed the seas in those early days were often only a little better off than the Roman galley slaves had been. Some of them were petty criminals who had been ordered to serve their sentences on the high seas. Others were poor young boys who had a keen sense of adventure, or who had no other way to make a living. The lives of the crew members were completely controlled by the ship's captain, who was often a cruel tyrant. Their days were filled with long hours of boredom, followed by longer hours of grueling work. Living quarters were crowded and unsanitary, and the food unappetizing. The scurvy that resulted from the poor diet caused teeth to loosen and skin to break out in sores.

In spite of the hardships, or maybe because of them, the relationship between a veteran sailor and his ship was often a close one. The vessel was his home, and the other crew members were his family. He knew his craft's weaknesses, as well as its strengths. He could usually ride it through the heaviest of seas and coax the last bit of mileage out of every stray breeze.

A skilled seafarer could predict trouble before it happened and was able to mend any part of his ship. If a mast cracked, and there was no spare, the ship's carpenter made a serviceable one from its splinters. If the steering gear needed repair, it was fixed at least well enough to get to the nearest port. On a bad voyage, a ship might need an entire new set of sails. The ship's sailmaker could handle the job, although just one of the heavy canvas sheets might require as many as two million stitches.

EVEN THE BEST designed sailing ships depended upon the whims of the wind and the course of the currents to carry them on their roundabout routes. During the 1800s the development of steam power offered hope for a new sort of freedom for seafarers—freedom to choose more direct routes, freedom to proceed whether or not there was a favorable wind, freedom from the back-breaking work of hauling sail.

The first steamship was launched in 1787, but it wasn't until Robert Fulton built his *Clermont* that this idea was taken seriously by many investors. Even so, most of the men and women who lined the shores of the Hudson River in 1807 to watch the progress of "Fulton's Folly" had come to scoff, not to praise it. One farmer didn't stay for long. After seeing the great paddle wheels dipping into the water and the furnaces belching smoke, he ran home, convinced that he had seen the devil going to Albany in a sawmill.

Meanwhile, boiler hissing and pistons groaning, the *Clermont* was moving up the Hudson River at a steady five miles an hour. The successful completion of its journey convinced at least a few far-sighted people that steam power was the answer to many of the problems of ocean travel. The British were the first to prove their faith in steam. In 1838, their *Great Western* plowed across the Atlantic at more than eight knots.

"How this glorious steamer wallops and gallops and flounders along," declared an enthusiastic passenger. "She goes like mad . . . puffing like a porpoise, breasting the waves like a seahorse, and at times skimming the surface like a bird."

The age of steam had arrived, but most ocean-going vessels still clung to their masts and sails. Steam engines could, and often did, break down. For many years, the engines were fired up only when the vessel was entering or leaving a harbor and in times of calm. When the swift clipper ships were developed in the mid 1830s, these many-sailed "greyhounds of the sea" soon dominated the transatlantic passenger trade and also the route from Britain to Australia. Running under full sail, with a brisk wind at her back, a clipper ship could outrace any chugging steamship. The uneven competition continued until the 1860s when improvements in the steamship enabled it to outrun a clipper in all kinds of weather.

Sailors became as familiar with the thumping, throbbing, and pounding of the new steam-powered ships as they had been with the creaking, groaning, and flapping of the sailing vessels. Their repair jobs on the engine were often only patchwork, and the ship might drift for several weeks while the work was being done, but the crew usually managed to bring it into port.

There were times when seafarers showed an incredible amount of bravery and ingenuity. In 1900, a British tramp steamer lost her propeller during a gale. Rising to the emergency, the crew turned the ship's head into the wind and dropped anchor. The sailors then flooded the forward holds to raise the stern. With the ship rolling and lurching in this unnatural position, the spare propeller was swung outboard by the cargo winches. Carefully, all the while fighting the raging sea, the crew guided the propeller onto a wildly gyrating tail shaft and secured it with a locknut.

The entire job took several days and was supervised by the ship's captain, who was suspended over the treacherous waters in a bo'sun's chair. Again and again, the motion of the ship swung him high into the air, then plunged him into the crest of a giant wave.

"It was just part of a day's work," he said when he was later awarded a medal for bravery.

As THE DECADES passed, the variety of vessels increased. There were different kinds of merchant ships, trawlers, and pilot boats. Graceful sleek craft raced in regattas, and squat versatile little tugs spent their days as workhorses. When World War I was declared, powerful and lethal submarines, battleships, and destroyers went into action. After the war, aircraft carriers joined the fleet.

The luxurious passenger liner came into its own in the early 1920s. The Atlantic crossing was advertised as the "voyage of voyages," and several ships were declared to be the best and fastest in the world.

The publicity that attracted people to these liners stressed the grand ballrooms, shops, swimming pools, theaters, and elegant dining rooms. It didn't mention the fact that the Atlantic Ocean was still the roughest and most dangerous of seas, especially in winter. Aside from the agony of seasickness, there was the constant and very real danger of a ship sinking. A lady who hadn't been prepared for the high waves and the pitching and lunging of the deck under her feet became nervous. She asked the captain how far from land they were.

"Not far—only about three miles," he replied.

Reassured by his words, the lady went to her cabin. She wouldn't have slept nearly so well if she had known the land the captain had referred to was three miles *beneath* the ship.

The author of a guide entitled *Ocean Notes for Ladies* advised her readers to dress in their best gowns at all times. By doing so, they would be assured a decent burial if they were washed up on shore after being drowned at sea. The bodies of the poorly clad steerage passengers no doubt ended up in some potter's field.

DURING THE 1930s and 1940s millions of tons of food,

manufactured goods, oil, and raw materials had to be kept flowing from continent to continent. The emphasis on ship design shifted to merchant and other working ships that could transport goods ever more quickly and safely. The 1950s brought nuclear power to the shipping industry. The endurance of a nuclear-driven vessel is limited only by the endurance of its crew. The nuclear-powered freighter *Savannah* can travel for three and-a-half years without refueling. Nearly a century and-a-half ago, the original *Savannah,* which was a steamship, could chug along for only four days before she had to stop for fuel.

The United States Navy joined the nuclear age on January 17, 1955. With the words, "Now underway with nuclear power," the commander of the *Nautilus,* the world's first atomic-powered submarine, announced a successful launching. The words sent a thrill of pride through everyone who had played a part in the vessel's development. Their ship could travel one hundred thousand miles without stopping to take on fuel. It had a life support system that would enable its crew to survive for months without resurfacing. Like a ghostly steel whale a city block in length, it could cruise silent and unseen under polar icepacks and past tropical reefs.

Following close upon the heels of this remarkable ship, atomic-powered aircraft carriers appeared. Nuclear icebreakers rammed and battered their way through eight-foot floes, thus clearing the water for Arctic shipping.

The need for transporting large amounts of oil led to the development of the supertanker, which is currently one of the world's largest man-made structures. Some of them can hold one hundred tennis courts on their decks, and their bows kick up a twelve foot breaking wave. These ships are fully automated, so despite their size, they need a crew of only thirty or forty to operate them.

Supertankers have solved many transportation problems, but they have created many more. It takes them a mile to turn around unaided, and almost four miles to stop from cruising speed. Their gigantic hulls have very little "give," so they are vulnerable to the battering effect of the sea. If repairs are needed, they must make their way to a shipyard that's equipped to handle them. There's no such thing as patchwork repair, or "making do" on board a supertanker.

Currently, there are only about fifty ports in the entire world that are big enough and deep enough to dock a supertanker. In most locations, the great ships have to anchor far offshore, a situation that leads to difficulties and dangers in unloading and loading their cargoes.

The explosive nature of many cargoes makes the loading and unloading of any modern merchant ship a tricky business. Several years ago, the *Sansania* docked in the Port of Long Beach in southern California. It was carrying naphtha, a highly volatile form of petroleum.

As the cargo was being unloaded, a cloud of vapor formed above the ship. Still, everything went well until the ship's telephone rang, activating a loud signal bell on deck. The bell apparently set off an explosion that lifted off the ship's entire superstructure. Some of the crew members were killed, and pieces of metal were later found four miles away from the site of the accident.

Because of a concern over the safety of ships and crews, experts are studying ways to lessen the dangers that exist both in port and at sea. They have created miniature "oceans," complete with a jetty, deep sea mooring and navigational buoys, and anchorages. Various sea conditions are produced by wave-making machines and the responses of model ships to these "seas." Meanwhile, other people are looking for better ways to handle toxic and explosive cargoes that are commonly carried by ships.

As THE COST of oil increases, and the supply diminishes, superships may become impractical. Already, there is talk about returning to some extent to the age of sail. The sailing ship of the future will probably be totally automated and computerized. Its sails will roll out from the masts on stainless steel tracks, and they will be set by one man pressing a few buttons, instead of an entire crew scrambling and climbing and straining at ropes. Nevertheless, the modern sailor may still feel, as did a seafarer of the 1880s, that a "full-rigged ship under sail is the finest sight in the world." There will always be something about the "whomp" of a fresh wind catching the canvas that is more satisfying than the silent efficiency of an automated engine.

3

Parting the Curtain of the Sea

AS SEAFARERS built better ships and ventured further and further abroad on unknown seas, they continued to while away the hours by telling stories about phantom ships and sea monsters; about mermaids who lured men into the shoals of uncharted islands; and about the distant edge of the earth where the waters boil and from which no one ever returned.

At the same time, using only the crudest instruments and their own senses, they began to build a body of factual knowledge about the seas over which they sailed. The first sailors noticed that the water along the shoreline rolled in and out twice a day without fail. They wondered, with a feeling of awe, if this rhythm was tied with a greater rhythm throughout the universe.

Later they found that they could use the tides to carry them safely away from shore and back again. When they discovered the swift ocean currents, they used them to their advantage also.

Some of the first recorded measurements of water depths were made five thousand years ago by the Egyptians who sailed along the Nile. Eventually, as vessels made their away along continental shorelines, their captains often used a weight tied to a long rope to find out how deep the water was. By "plumbing the depths" in this manner, they could avoid the shallows and underwater outcroppings that could damage or ground their ships. This same technique was used by the river pilots of the Mississippi River during the 1800s. The measurements provided a reliable gauge of the river's shifting depth.

The longer the voyage, the more intensely sailors had to concentrate on the ocean's behavior. During the fifteenth and sixteenth centuries European explorers kept detailed journals about tides, currents, water temperatures and the sea creatures they observed. They mapped and charted their routes and found that all of the parts of the "world ocean" were connected by straits and passages. The broad expanse of water was no longer thought of as a barrier, but as a relatively smooth and easily traveled avenue to other lands and peoples.

Magellan, who led the first voyage around the world, was one of the many seafarers who were curious about the depths of the ocean floor. In 1520, he tied two 600-foot ropes together, attached a lead weight to one end, and lowered the entire length into the Pacific Ocean. To his great disappointment, he was "off sounds," that is, the weight failed to touch bottom.

After Magellan's attempt, no known deep sea soundings were taken for 250 years. The first sucessful one occurred in 1773, in the sea between Norway and Iceland. The weight touched bottom at 683 fathoms, or almost 4,100 feet. When it was brought to the surface, it was covered with a fine blue clay. The men who handled this ordinary-looking substance were aware that they were touching something that had never before been felt or seen by a human being.

About fifty years later, another ship's line was found to be "on sounds" at one thousand fathoms. That time the weight brought up some greenish mud, and the net that was attached to it contained some worms and a starfish. This discovery startled most scientists. How can life exist that deep in the ocean? they asked. How can any creature survive with no sunlight, no air, and in such a cold temperature? How can it withstand pressures that are capable of buckling the most durable of metals?

It was soon realized that these first successful soundings were far from accurate. While the line was being lowered, the ship was drifting. By the time the weight touched the sea floor, the line was no longer vertical, but at an angle. Thus, the measurement was always much greater than it should have been.

In 1840, that problem was partially solved. Two ship's longboats were tied together. One of them held four miles of rope on a reel. In the other were two oarsmen, who had been given the backbreaking job of rowing constantly against the wind and the current. Through their efforts, the two boats could be held "on station," or in one location and the lowered line would be kept perpendicular to the surface.

During the many hours that it often took for the weight to reach the bottom, these men had to ignore the cold, their hunger pangs, and the increasing ache in their arms and backs. In a limited way, their agony was worthwhile. Some of those measurements were extremely accurate. In many cases, however, they were very wrong. At that time, no one knew about the submarine currents that exist far below the surface of the water. The force of their flow was usually strong enough to make a rope drift, no matter how steadily it was being held at the surface.

It was plain to see that if each sounding was going to take so long and involve so much grueling effort, only a small part of the ocean floor would ever be measured. Over the next few years, faster and more dependable sounding devices were invented. Improved dredging instruments also brought to light an increasing number of odd sea creatures, a few of which were "living fossils" that had been thought to be extinct.

THE VOYAGE of the HMS *Beagle*, which lasted from 1831 to 1836, contributed a great store of information to oceanic map makers. Although it was primarily a surveying expedition, it was on this long cruise that the naturalist Charles Darwin gathered the geological and biological data that led to his theory of evolution.

The next giant step in the study of the world ocean was taken after the telegraph was invented in 1837. European and American businessmen were eager to use this speedy method of communication, so they enthusiastically supported the plans to lay a transatlantic cable. The success of the project depended upon finding the answers to some questions about the proposed cable route. What sort of water movement exists there? How is the ocean floor contoured? Are there any animals inhabiting the region, and if so, what kind? Will they chew the cable's covering? Or will they deposit their wastes upon it, thus causing it to rot?

Biologists, botanists, zoologists and oceanographers went to work to find the answers to these and other questions. They took soundings at regular intervals all along the route. They analyzed sediments from the sea floor. They made charts of the currents.

One weight that had been left on the sea floor for an hour came back to the surface with thirteen starfish clinging to it. The captain was delighted at what he considered proof that abundant life existed at great depths.

"The sea has sent forth its long coveted message," he wrote in his log that night.

Some of the scientists weren't certain that the starfish had actually lived on the sea floor. "Couldn't they have clung to the weight as it passed through the water at a much higher level?" they asked.

In 1860, their doubts were partially resolved. At that time, a submerged cable broke and was brought up for repair. Several sea creatures were clinging to it, one of which was a coral which could only have grown into place as the cable lay on the ocean bottom.

By the time the Atlantic cable project was completed, this particular section of the sea floor had been mapped in a detail that had never before been attempted. A vague picture of deep coastal trenches and a central plateau began to emerge. Until that time, it had been assumed that the sea floor sloped gradually down to its lowest point about halfway between the two continents, then climbed back up to meet the opposite shore. It now appeared that the middle of the ocean floor wasn't its lowest point, but perhaps its highest.

As usually happens in a scientific investigation, these discoveries led to a host of questions that cried out to be answered. What does the presence of these undersea mountains and valleys mean? Are they part of the lost continent of Atlantis? Or are they part of a newly emerging landform, one which will appear above the surface of the sea millions of years hence?

To solve these and other mysteries, the world's major seafaring countries joined in the search for more facts. The "space race" of the nineteenth century began, but instead of competing, the participants cooperated with each other.

In 1872, England launched the HMS *Challenger*, which was to "investigate the condition of the Deep Sea throughout the entire Oceanic Basin." An old naval sailing ship with auxiliary steam power, it was an unlikely vessel to perform such a monumental task. Besides being clumsy and slow, it had an alarming tendency to pitch and roll much more than most ships.

As the *Challenger* groped its way across unknown seas using inaccurate and partially blank maps, the members of its crew often had to call upon all of their skills of seamanship. In the Antarctic, the ship was caught amid icebergs that were drifting in the gale force wind. By artfully navigating between two icebergs that were used as a windbreak, the sailors were able to escape the precarious situation.

Each time the *Challenger* had to "take a station" to make a depth sound, the sails had to be shortened, and the ship brought into the wind. Twin propellers, which were driven by the steam engines, were used to keep the craft steady in the water. Steam-powered winches creaked and groaned as the lines went slowly over the sides.

Each line could hold a variety of primitive gear. The nets that gathered the plankton and other forms of small marine life were muslin or silk bags attached to iron rings one foot in diameter. Water samples were collected in glass bottles that could be opened underwater by the men on the ship. Mercury recording thermometers measured the water temperature. These readings were far from accurate, and the instruments were often broken when they struck the side of the ship.

From the time the *Challenger* furled its sails to take a station to the time when the last foot of line was wound laboriously back onto the reel, two days sometimes passed. Stations were often held in the fury of a raging blizzard and in below-freezing temperatures. As the scientists waited for the dredging to be completed, they suffered agonies of frustration and uncertainty. Would the steam engine break down at a crucial moment? Or would the winches jam? What if the nets came up empty? Even the ship's parrot seemed uneasy. "What?" he echoed. "Two thousand fathoms and no bottom?"

The *Challenger*'s ordinary seamen had little understanding of the reasons behind the grueling search for bits of coral and shell and mud and clay. They knew only that they had to work long hours under the worst of conditions and that they were often on the receiving end of the scientists' anger when something went wrong. It was hard for them to understand how a man's happiness could depend to such a great extent on some slimy bit of muck, or a particularly unattractive sea worm.

THE HMS *Challenger* returned to England in May of 1876. During the three and one-half years of its voyage, it had logged almost sixty-nine thousand miles and had taken over 360 stations, having stopped on the average every 200 miles.

The scientists were jubilant. They had measured the structure of the ocean all around the world and had found it to be marvelously complicated. They had analyzed the chemical composition of hundreds of water samples and saw that the sea contained many of the earth's minerals in solution. They had dredged up thousands of species of animal life that were new to the scientific world. Their findings had indicated that life exists in a myriad of forms in all of the measured depths of the sea. They had dredged up strange, charred-looking rocks that ranged in size from cinders to potatoes. These were put on display in the British Museum, but otherwise received little attention.

The *Challenger*'s scientists had felt the thrill of handling ooze that had lain hidden under thousands of fathoms of water for millions of years. This sediment was composed largely of the corpses of microscopic creatures that constantly filter like an endless snowfall to the ocean floor.

One of the most awesome results of the *Challenger*'s voyage was the partial uncovering of the hidden world of the ocean floor. This world contained a tantalizing mixture of tall mountain ranges, wide plateaus, and deep, plunging trenches. Far from being flat and uninteresting, the sea floor apparently rivaled the continents in the majesty of its landforms.

The *Challenger* had collected so much information that it took scores of experts twenty years to sort it out. The data eventually filled fifty thick volumes, and the study of the material kept the oceanographers of the world busy for many decades.

"Never did one expedition cost so little and produce such momentous results for human knowledge," wrote a scientist. "Indeed, the science of oceanography began the day the *Challenger* was launched."

Despite such words of praise, the *Challenger*'s voyage was only the beginning. The experts who had been on the vessel readily admitted that they had been "like blind men, groping about in the water with white sticks three and-a-half miles long." Many of the mysteries that were found to exist at that time remain unsolved today.

4

The Restless Waters of the World Ocean

CENTURIES BEFORE there was any serious study of the world ocean, sailors had found that the best way to get from one place to another was seldom the shortest way. As the currents carried them first in one direction, then in another, it was often only through trial and error that they finally reached their destination. Information about good routes was passed on by word of mouth and by crude charts. The sailing directions given to Vasco Da Gama in 1497 provided him with the best possible passage between Europe and the Cape of Good Hope.

Every sea captain kept notes about daily wind and weather conditions, but sometimes what they learned wasn't put to practical use. When Benjamin Franklin was the Deputy Postmaster of the British Colonies in North America, he received frequent complaints from merchants about the slow delivery of mail from England. For some reason, the British mail packets took two weeks longer to make the trip to the colonies than they did to make the return trip. After talking to colonial seafarers and fishermen, Franklin realized that the westward bound British ships were bucking a strong eastward flowing current.

"We've told the English captains about that current," said several of the fishermen, "but they don't pay any attention to us."

Franklin himself didn't have much more success when he told the British sailors to sail to the side of the current. He was, after all, only an "upstart colonist."

The snub made Franklin angry enough to dip into his own personal savings to pay for a study of the North Atlantic currents. By collecting information from the logs of numerous ships that regularly plied the route between England and the colonies, he was able to publish a chart in which the Gulf Stream was clearly shown. His work convinced even the most stubborn of Britishers that this stream did indeed exist.

Franklin's simplified chart inspired other such studies, because it had been proved that a knowledge of the currents could save businessmen a lot of money. Not long after his study, a current similar to the Gulf Stream was charted in the Pacific Ocean.

Despite such progress, it wasn't until the late 1800s

that anyone made a serious effort to put together all the scattered bits and pieces of information about the ocean currents. It was then that Lieutenant Matthew Maury of the United States Navy began pouring through thousands of old ship's logs. From the data that he gleaned, he compiled a wind and current chart of the North Atlantic that was much more complete than Franklin's had been. Maury then went through many more thousands of logs and put together the first global wind and current chart.

Only one year after the publishing of these charts, ship owners had already saved millions of dollars by following Maury's sailing directions. For one trip alone—that from the British Isles to California—thirty days were cut from the schedule. Maury also saved many lives by showing mariners how to avoid shoals and dangerous rocks.

Maury's charts and succeeding studies clearly showed that the major currents take the form of great "gyres," which look like giant whirlpools, or constantly revolving wheels that fill entire ocean basins. In the northern hemisphere, the clockwise rotation of the gyres causes water to pile up on the eastern shores of continents. In the southern hemisphere, the counterclockwise gyre rotation causes water to pile up on western shores. In each case, the piling up forms currents that are swift and concentrated.

There are five major gyres. They are located in the North and South Atlantic, the North and South Pacific, and the Indian Ocean. Four of these gyres aren't affected by seasonal weather changes. The exception is the Indian Ocean gyre. Here, from April to October, the powerful monsoon winds reverse the flow of the upper half of the gyre. Meanwhile, the southern half proceeds as usual.

The much-studied Gulf Stream, the so-called "Ocean River," is the swiftest flowing part of the North Atlantic gyre. It rides past Miami, Florida, at speeds up to five miles an hour, almost twice as fast as an average surface current. Like a mighty flood fifty miles wide and fifteen hundred feet deep, it sweeps four billion tons of water each second toward the shore of North Carolina, then veers northeastward. Near Newfoundland, its warm blue mass runs into the icy waters of the Labrador Current. The collision thrusts long, cold fingers into the Gulf Stream flood and produces billowing clouds of fog. Much of the heat carried by the Gulf Stream is dissipated along the western shores of Europe.

Maury was fascinated by the Gulf Stream, the size and power of which makes the largest continental rivers look like meandering meadow streams. "The Gulf of Mexico is its fountain," he wrote, "and its mouth is in the Arctic Sea. There is in the world no other such majestic flow of water. Its current is more rapid than that of the Mississippi or the Amazon, and its volume is a thousand times greater . . . ."

The Sargasso Sea, the "Graveyard of Ships," is the hub of the Atlantic gyre, the point about which it turns.

It revolves sluggishly and is able to support only a small amount of life. Its predominant plant life is a brown seaweed, the fronds of which are inhabited by miniature snails and crabs, and tiny fish. Atop the seaweed runs the waterwalker, a spider with threadlike legs. These little creatures are a far cry from the monsters that were envisioned and so greatly feared by the early explorers.

The northern Pacific has its own ocean river in the form of the dark blue Kuroshio Current. It begins near the Philippine Islands and during the course of its long journey meets the cold currents that flow down from the north. The fogs and storms caused by their mingling rival those found near Newfoundland.

In addition to the main gyre, the icy circumpolar currents sweep around both poles, then drain back into the central basin. The Antarctic Circumpolar Current is the only one in the world that isn't obstructed by large landforms. Its steady, uninterrupted flow and the strong, squally winds of the region have caused seafarers to name these latitudes the "Roarin' Forties," "Howlin' Fifties," and "Screamin' Sixties."

MANY OF the early oceanographers had a vague awareness that there might be some sluggish, insignificant water movement in the depth of the sea, but they believed that those currents were unimportant. It wasn't until after World War II that a serious study of submarine currents began. Some of the discoveries were accidental, such as the one that was made in 1951. At that time, a research vessel in the South Pacific was testing new fishing techniques. Its crew let down cables that were several miles long and that had shorter lines hanging from them. The observers expected to see this bulky equipment drift off to the west with the prevailing current. They were astonished to see it drift eastward instead. The lines had been caught up in a fast running and extensive undersea current that runs counter to the surface current. Its total length is still not known.

Within a few years another current was found running under and in an opposite direction to the Gulf Stream. Recently, an intermittent, unpredictable high speed current was found flowing at the bottom of the North Atlantic.

The cause of some of these deep currents is known. During the trip from the equator to the poles, ocean water loses its heat to the atmosphere. When it combines with the runoff from the melting of ice, it becomes cold and heavy, then sinks. When it reaches the sea floor, it displaces the water that's already there, thus setting up submarine currents. These currents carry the cold water back to the equator where it rises, is warmed, and starts the long trip back to the poles.

Many of the deep currents travel at high speeds, while others just dribble along. Up to fifteen hundred years may pass before the slower-moving water will again be exposed to the rays of the sun. But as thousands of years pass, every droplet in the world ocean will sink and rise, rise and sink, and circle the globe again and again. In this way, any liquid or dissolved substance that is put into the sea will eventually be distributed by the currents equally throughout every ocean region on our planet. For this reason, the ocean floor can no longer be considered a safe dumping ground for our poisonous wastes.

IT'S APPARENT that the restless waters of the world ocean present a complex and intriguing puzzle that men have only begun to solve. Scientists do know that the study of ocean currents is closely related to the study of air movements. Within each of them, warm, light masses rise, while cooled, heavy masses sink. They are both affected by land forms and by the earth's population. They are closely coupled, because the wind furnishes much of the power that stirs seas into action.

A map of global air and water currents shows order, regularity, and smooth, sweeping motions as they flow in their endless cycles. To the captain of a small ship, however, the reality is much different. In even the calmest of weather, the water's movement is often turbulent. Whenever cool water meets warm, or salty water meets fresh, or any high density water meets any low density water, smaller, conflicting currents may be set up. Cells and eddies may swirl and mingle, meandering little branches of currents may flow off to one side, countercurrents may overlap like shingles on a roof, and upwellings may bring great volumes of water rushing upward from the sea floor. During the Gulf Stream's eastward swing, it breaks up into separate bands and swirls, some of which form independently moving rings. Some of these rings are over sixty miles in diameter and may last up to a year before they flow back into the main gyre.

Some of the best fishing in the world is often found in the most chaotic waters. Off the Grand Banks of Newfoundland, where the Gulf Stream meets the Labrador Current, millions of fish feed upon the nutrients that are brought up from the seafloor by the upwelling of water. Upwelling often occurs when a strong, steady wind blows parallel to a coast, causing a displacement of surface water. It's also caused by currents interacting with the sea floor in certain ways. In either case, the surface water is replaced by water from lower ocean layers. The upwelling water is colder and contains less oxygen than the displaced water, but it's rich in the phosphates and nitrates that serve as nutrients for fish.

The upwelling that occurs near the Grand Banks results in a flourishing fishing industry, one that has produced some of the finest sailors the world has ever known. Unfortunately, many of them have paid for their rich harvest with their lives.

Fishermen must be prepared to cope with the malicious tricks that nature occasionally plays on them. In 1882, an extraordinary number of icebergs appeared in the Labrador Current. As they melted, the additional water strengthened the current's flow, and the Gulf Stream was pushed southward. Since the front between the two water masses was now located in deep water, some distance from the shoreline, there was no longer any upwelling from the sea floor. The fish moved on in search of nourishment. Until the front returned to its usual position, the fishermen had to hunt for the new fishing grounds, and their nets often came up empty.

There are places throughout the world in which current-related disasters occur regularly. The coast of Africa is one of them. Another is coastal Peru. Here, a wind usually hits the mountains and is deflected northward to run parallel to the coast. The surface water that's carried along by the wind is known as the Humboldt Current. Cold water upwells to replace the water that's swept away, and the native fish and birds thrive on the nutrients from the subsurface water. Indirectly, so do the many Peruvians who make their living by fishing and by processing the fertilizer that's made from bird droppings.

This happy state of affairs is disturbed once every few years when widespread changes occur throughout the southern Pacific gyre. These changes are related to a weakening of the equatorial trade winds. This weakening allows a tropical air flow to bring an unusual amount of warm water southward. A warm sluggish water movement interferes with the cold, swift Humboldt Current, and the upwelling ceases. As the birds and fish leave the area in search of food, the jobs of the people who depend upon them for a living disappear also.

The search for knowledge about the ocean's current system will continue for many more decades. Perhaps it will never end. Oceanographers, people who work for governmental agencies, and ship captains are constantly gathering new facts all over the world. Improved current meters and drifting buoys measure not only the speed and direction of currents, but also their salinity, temperature, oxygen content, and the amount of sunlight that filters through the surface water. A particular current's plant and animal life are often closely monitored.

The launching of Seasat-I extended the search for facts about the ocean into outer space. This satellite, and others, enable oceanographers to study an entire ocean basin at one time. Their instruments relay measurements of incredible accuracy. One of them is able to sense the exact height that a swiftly moving current rises above the surface of the sea that surrounds it.

FROM EAST TO WEST, from north to south, from the sandy continental shelves to the dark abyssal plains, currents are circulating. Because of their movements, the saltiest of water becomes less salty, the warmest of water becomes cool, and the coolest warm. Currents carry oxygen to the deepest parts of the sea and lift nutrients from the ocean floor to the surface. By carrying the ocean's stored heat, then releasing it along continental shores, these "arteries of the earth" temper harsh climates. They determine where vegetation will flourish and where deserts will spread. As a result, they have a great influence on where human beings will live and work.

The constant and dependable movement of the ocean's restless waters is as important to our earth as the movement of blood is to our bodies. Without that circulation, the pattern of life on our planet would quickly unravel and disintegrate. By continuing to study the currents, we'll be better equipped to hold that pattern together.

Many ships foundered during the dangerous journey through rough seas and high winds around Cape Horn at the tip of South America. The Whaling Museum, New Bedford, MA

For many years various forms of the outrigger canoe have been used by natives of the Pacific Islands. Hawaii Visitors Bureau

OVERLEAF: *Currents are constantly moving water around the world. This movement helps to equalize temperatures, cooling some regions and warming others. Without currents, many parts of the world would be uninhabitable.*

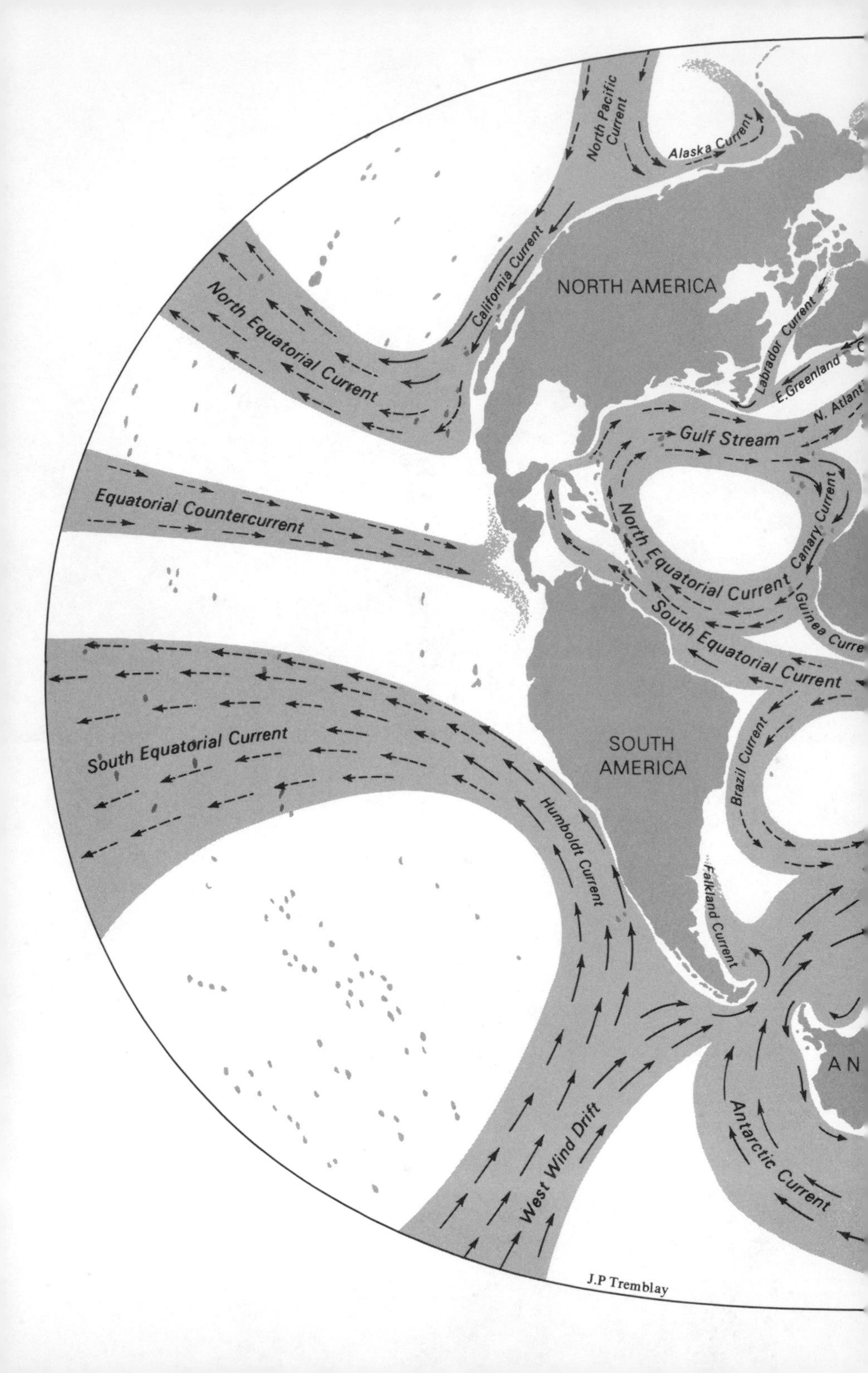

North Pacific Current
Alaska Current
California Current
North Equatorial Current
NORTH AMERICA
Labrador Current
E.Greenland
N. Atlant
Gulf Stream
Equatorial Countercurrent
North Equatorial Current
Canary Current
Guinea Curre
South Equatorial Current
South Equatorial Current
SOUTH AMERICA
Brazil Current
Humboldt Current
Falkland Current
West Wind Drift
Antarctic Current
AN
J.P Tremblay

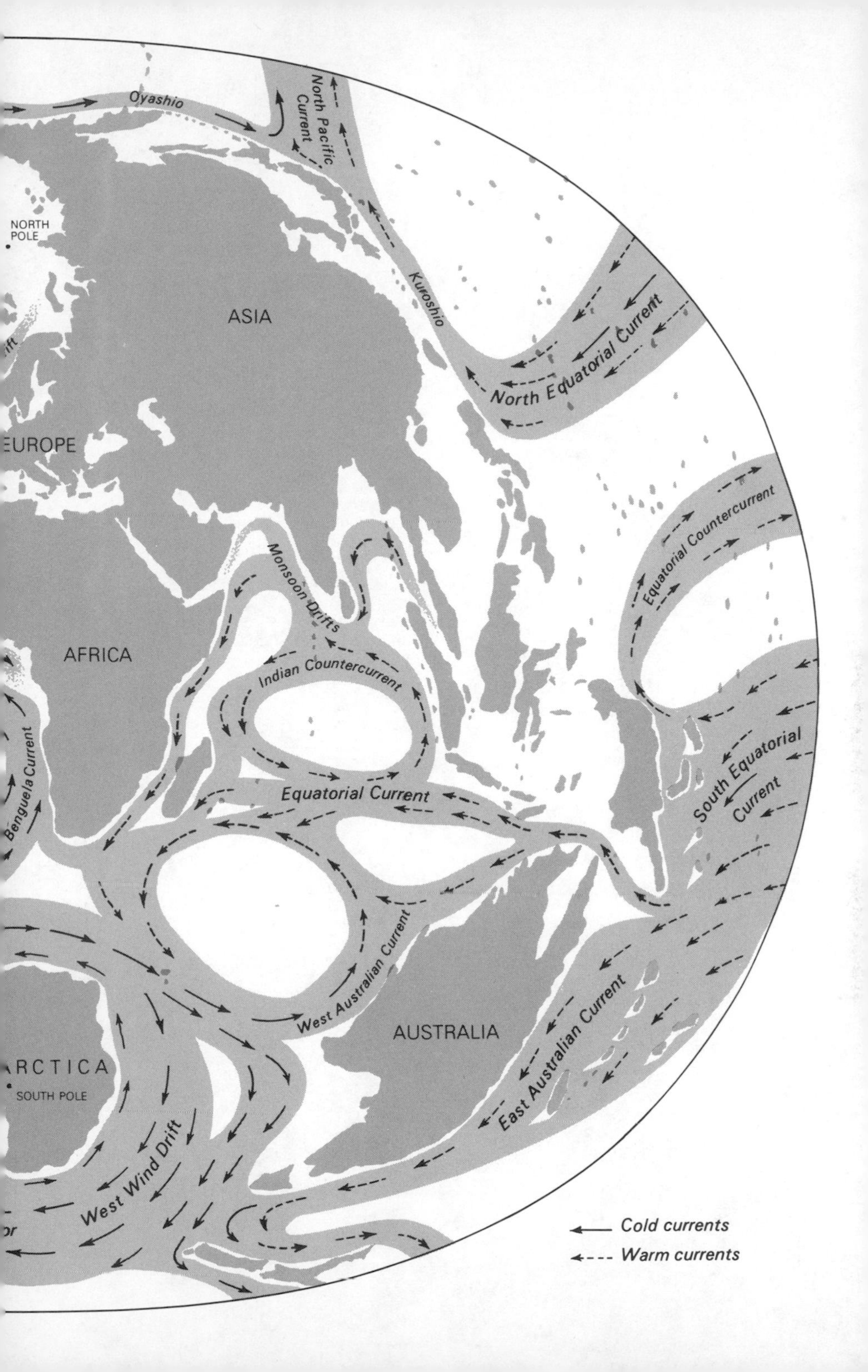

Oyashio
North Pacific Current
NORTH POLE
Kuroshio
ASIA
North Equatorial Current
EUROPE
Equatorial Countercurrent
Monsoon Drifts
AFRICA
Indian Countercurrent
Benguela Current
Equatorial Current
South Equatorial Current
West Australian Current
AUSTRALIA
East Australian Current
RCTICA
SOUTH POLE
West Wind Drift
Cold currents
Warm currents

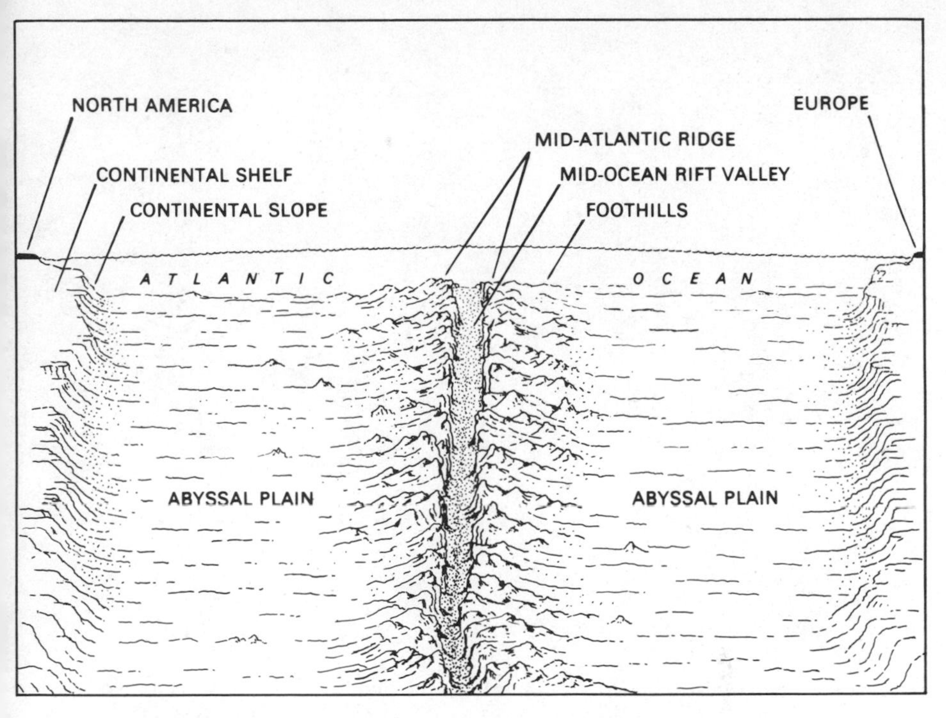

Sonar mapping has enabled us to visualize the profile of the ocean floor. Many of the features dwarf anything to be found on dry land.

Today's oceanographers have access to sophisticated equipment. This tripod is used to measure deep ocean currents. Jet Propulsion Lab

This topographic map of the ocean floor shows the winding Mid-Ocean Ridge.

A camera sled is launched for a survey of the ocean floor. Woods Hole Oceanographic Institution

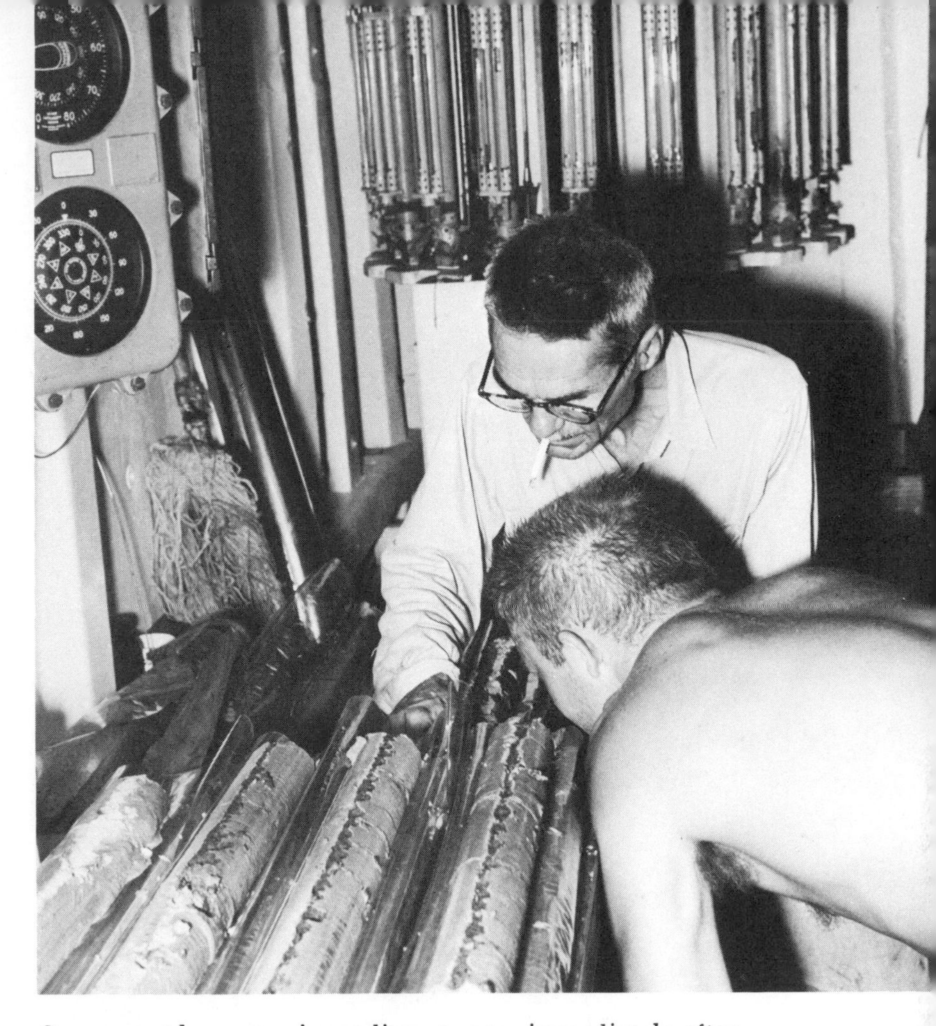

Oceanographers examine sediment cores immediately after recovery from the bottom of the sea. After the recovery, the cores will be stored under refrigeration for future study. Scripps Institution of Oceanography

OVERLEAF: *The longest chain of mountains in the world lies underwater. Known as the Mid-Ocean Ridge, its branches can be found in the Atlantic, the Pacific and the Indian Oceans.*
In the center of the Mid-Ocean Ridge is a rift into which molten rock from the earth's mantle erupts. It's within these volcanically active rifts that sections of the ocean floor are spreading apart and moving in opposite directions.

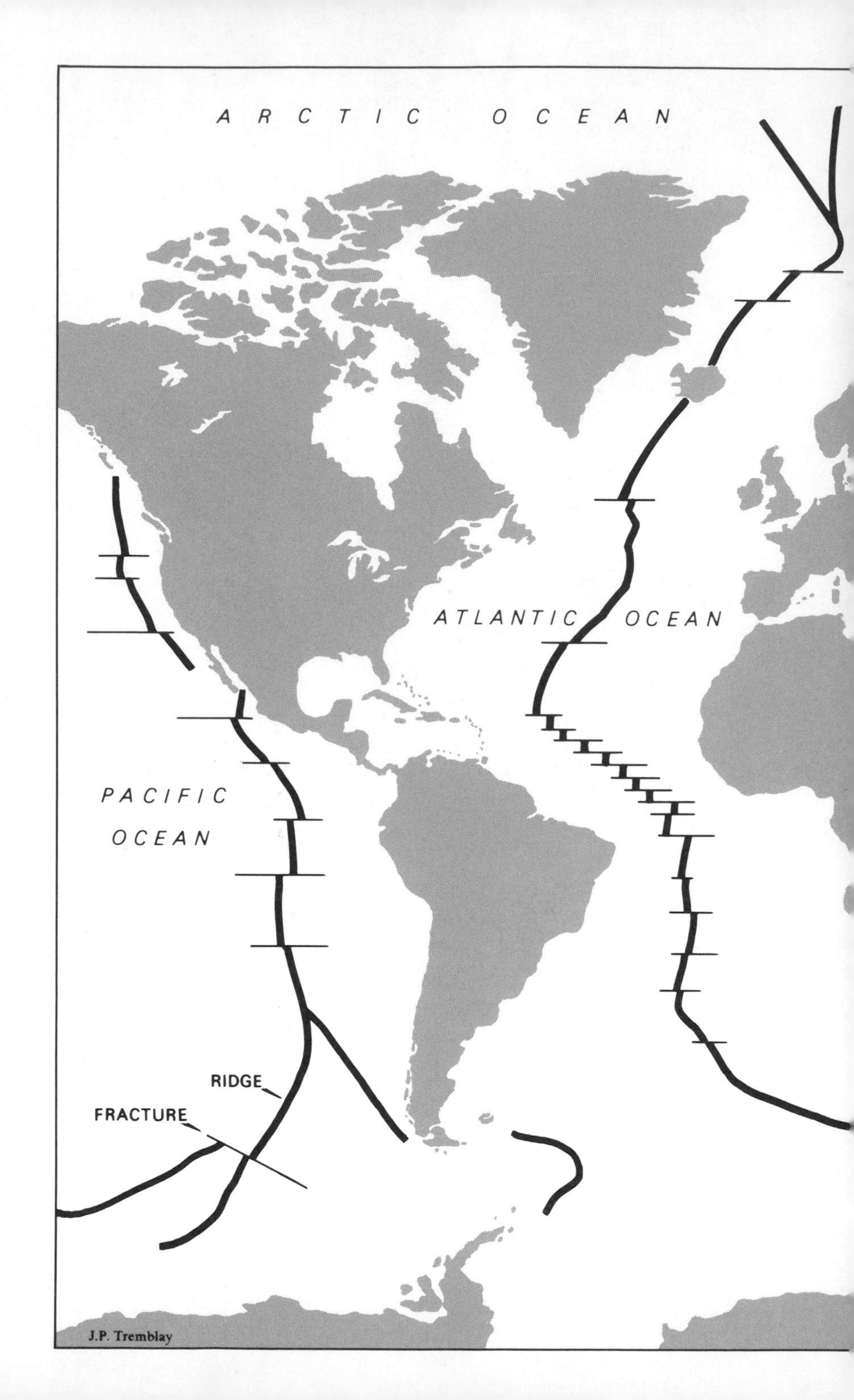

ARCTIC OCEAN
ATLANTIC OCEAN
PACIFIC OCEAN
RIDGE
FRACTURE
J.P. Tremblay

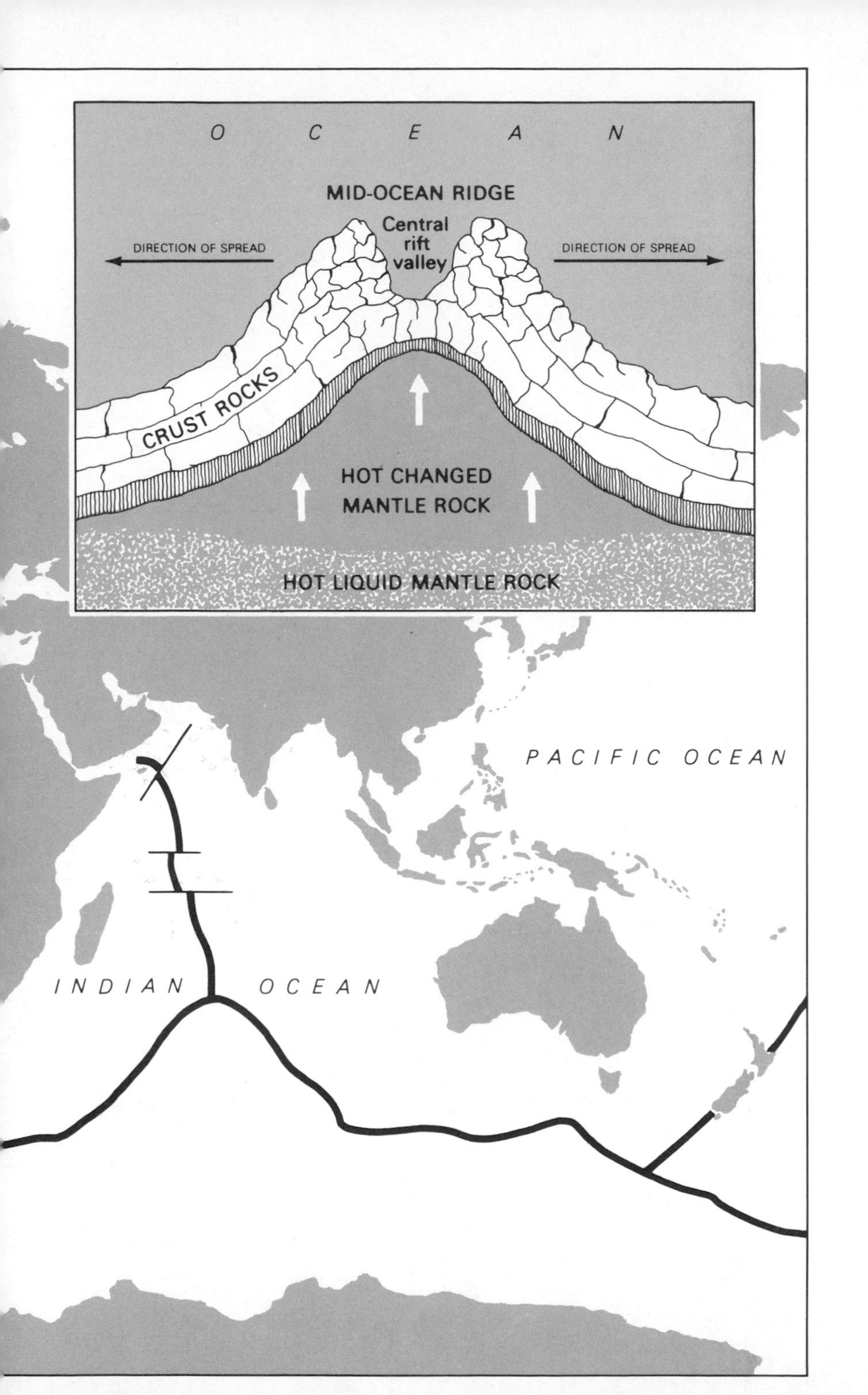
O C E A N
MID-OCEAN RIDGE
Central rift valley
DIRECTION OF SPREAD
DIRECTION OF SPREAD
CRUST ROCKS
HOT CHANGED MANTLE ROCK
HOT LIQUID MANTLE ROCK
PACIFIC OCEAN
INDIAN OCEAN

On board the research vessel Oceanus, *a student uses a computer to analyze data. Woods Hole Oceanographic Institution*

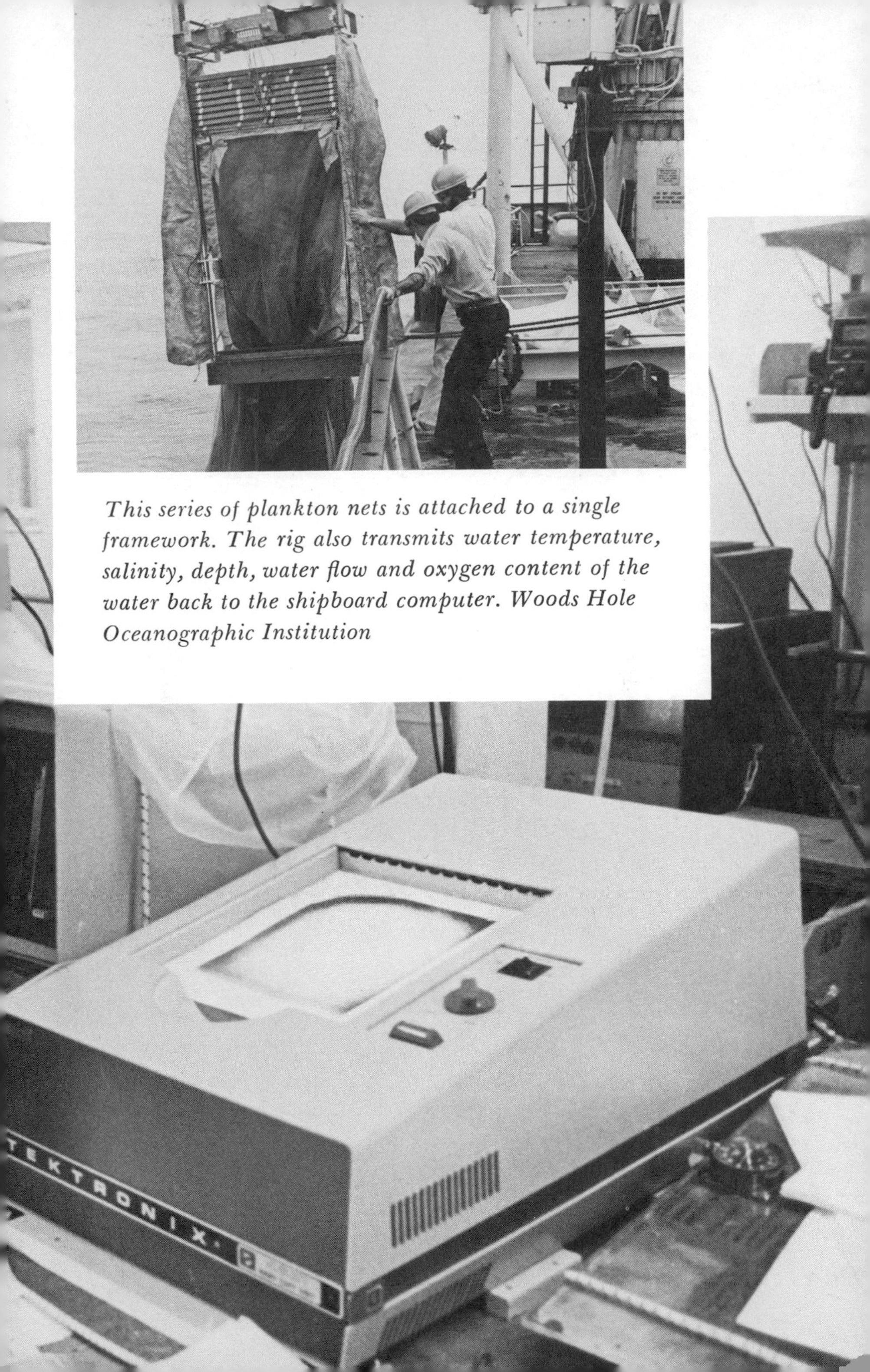

This series of plankton nets is attached to a single framework. The rig also transmits water temperature, salinity, depth, water flow and oxygen content of the water back to the shipboard computer. Woods Hole Oceanographic Institution

A wave crashes upon a beach, throwing up a frothy foam, the "dust of the sea." Hawaiian Visitors Bureau

Waves have carved this rugged piece of Hawaiian coastline. Hawaiian Visitors Bureau

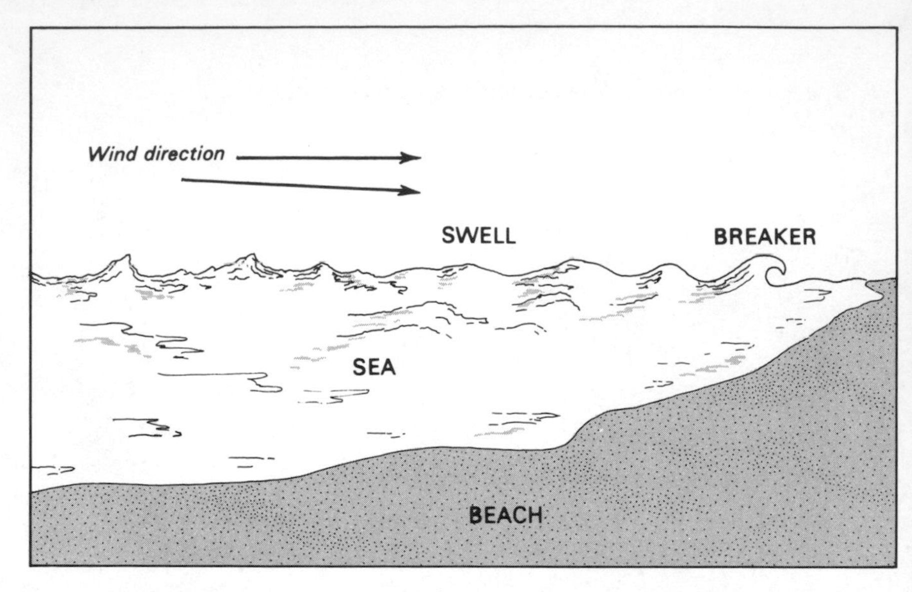

Winds cause waves on the ocean's surface. As a wave progresses, it moves energy, not water. Thus, an object caught in a wave won't move forward. After being carried by the circular motion of the water within the wave, it will end up back where it started. When the wave breaks upon the shore, however, its energy is expended. Then, an object caught in the wave may be thrown out onto the beach.

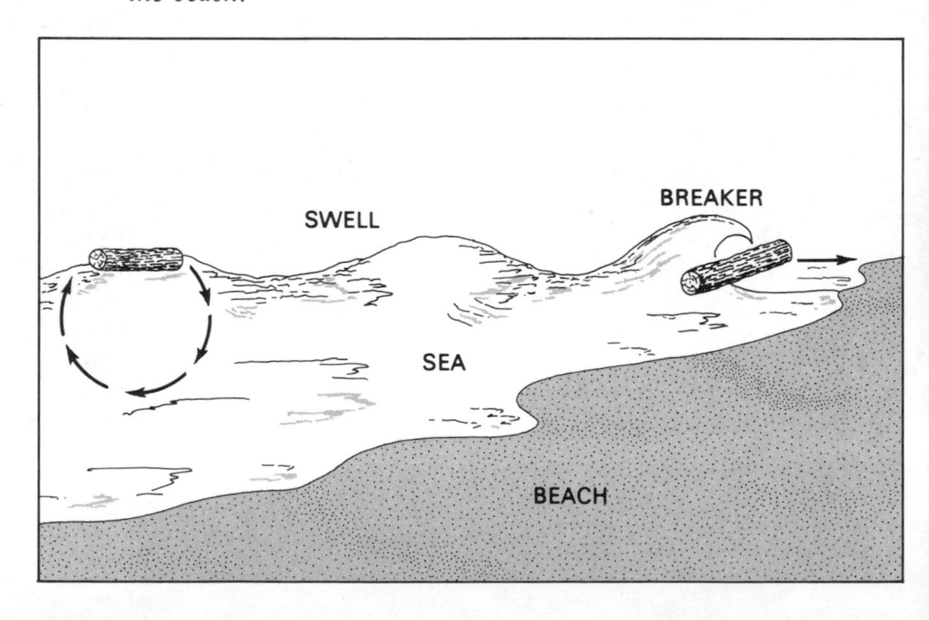

5

"A Tangled Jumble of Massive Peaks"

BECAUSE OF the voyage of the *Challenger* and Maury's charts of the currents, the science of oceanography took some giant steps forward. Nevertheless, as the twentieth century began, the topography of the ocean floor remained a fuzzy and incomplete picture in the minds of even the most knowledgeable scientists. Each sounding still took hours of grueling work, and the painfully obtained measurements were miles apart. Oceanic map makers longed for a way to fill in those gaps.

The technology to meet their needs was developed in response to two other big problems. The sinking of the *Titanic* in 1912 brought into sharp focus the necessity for more accurate and dependable iceberg detection.

Soon after that tragic accident, World War I broke out, and submarines prowled the Atlantic Ocean. Their crews needed better ways to communicate with the ships of their own country, plus a method by which they could detect the enemy's elusive craft. Sound transmission was seen as the technique that could do these jobs.

"Sonar," as the new technology came to be called, relies on sound impulses that are produced by an acoustic projector, a sort of loudspeaker. The sound waves emitted by the loudspeaker can travel long distances. When they strike an object, they bounce back, or echo. When the echoes are timed, they give an accurate measurement of the distance of the object. As they are further analyzed, they give information about the size, shape, and nature of the object.

When the war ended, oceanographers wasted no time in equipping their ocean-going laboratories with sonar. After beaming their "sound searchlights" downward, they watched a stylus move across a chart as their ship proceeded above the area to be mapped. Within seconds, sonar took soundings that would have previously taken years. The valleys and canyons, the hills and mountains, the awesome grandeur of a hidden world was suddenly there for all to see.

There were a few problems with the first sonar equipment. Inexperienced operators onboard submarines often mistook schools of fish for other submarines. On a research vessel, if a ship's cook opened or closed a heavy cupboard door, the vibration might produce a canyon where no canyon existed. Other common shipboard events resulted in a variety of other phantom landscapes. These "ghosts" disappeared as sonar was improved.

With the help of this remarkable new tool, scientists have been able to obtain an extremely accurate three-dimensional map of the world that had always been cloaked in darkness and mystery. This world has proved to be more remarkable and astonishing than anything to be found upon the continents.

Sonar soundings have shown how the sea washes onto the gently sloping margins of land that form the continental shelves, which range from a few miles to several hundred miles in width. Seaward from the shelves lie the highest bluffs on earth, the continental slopes. These slopes plunge three or four miles down to the ocean floor. Spectacular V-shaped canyons are carved into the faces of them. Off the coast of California, the submarine Monterey Canyon is a twisting, 100-mile-long gorge that is flanked by walls six thousand feet high and ten miles wide. There is some evidence that undersea landslides, or "turbidity currents," helped to carve these canyons by washing sharp-edged sand and gravel downward.

At the base of the slopes, the continental rise begins. Here a vast network of broad sediment fans extend onto the abyssal plains, the flatlands of the sea.

The abyssal plains are unlike any other region on earth or in the explored parts of outer space. Here the ocean floor is covered by fine muds, clay, sediments, and ooze, which is composed of shells and skeletons and dead plankton, which are very small, often microscopic, plants and animals. For mile after endless mile, these materials blanket the area, forming a smooth, almost uninterrupted surface. Occasionally a dome-shaped hill rises from these vast plains. Millions of years from now, these hills will also be covered by delicate sheets of sea floor sediment.

Here and there isolated peaks or clusters of peaks tower over the domed hills. These "seamounts" are actually volcanoes, some of which poke through the ocean's surface to form islands. Some seamounts have flat tops, and are called "guyots," after the Frenchman who first discovered them. It's believed that they used to be volcanic islands that had their tops sheared off by the cutting edge of breakers and surf. As millions of years passed, they sank back into the sea from which they had originally sprung.

Dominating all of the features of the sea floor is a series of massed peaks, the outlines of which are as raw and rugged as a shark's teeth. They form a mountain range that is higher and longer than any other range on our planet. For forty thousand miles it winds through the various ocean basins. Sometimes it reaches high enough to form islands such as the Azores in the Atlantic. In some places it's only a few miles wide, while in others it's more than a thousand. Its peaks fill the entire central third of the Atlantic Ocean basin, where it's known as the Mid-Ocean Ridge. Winding down the center of the range is a steep walled crack, or rift. In some places the rift is deep and wide enough to hold the Grand Canyon with plenty of room left over.

Taken in its entirety, this range is the single largest geologic and geographic feature in the world. Oceanographer Maurice Ewing described it as "millions of square miles of a tangled jumble of massive peaks, sawtoothed ridges, earthquake-shattered cliffs, plunging valleys, lava formations of every conceivable shape. . . ."

WHILE early oceanographers were mapping the sea floor, a young German named Alfred Wegener was studying the shapes of the continents. Like others before him, he was struck by the fact that they looked like scattered pieces of a giant jigsaw puzzle. It seemed, for instance, that the eastern shore of Africa could once have fitted neatly against the eastern shores of North and South America. Wegener became convinced that at one time in the distant past, they had actually been joined, then somehow drifted apart, and in 1912 he presented his ideas.

Most of Wegener's fellow scientists were outraged. "If we are to believe this hypothesis," one of them protested, "we must forget everything which has been learned in the past seventy years and start all over again."

At that time, Wegener had only his observations of

rock formations to prove his theory, but later, more evidence came in to support it. It was found that there is a constant flow of heat throughout the partially melted plastic-like layer of rock that lies directly under the earth's top crust. Scientists also found that the rift valley of the Mid-Ocean Rift is an "unhealing wound in the earth's crust," an area in which the sea floor is "bursting with volcanic activity." Parts of the rift are occasionally filled with molten material that has bubbled and spewed up from the earth's interior.

In 1960, the theory of sea-floor spreading exploded like a bomb on the scientific community. This idea caused more arguments than any theory since Darwin's theory of evolution. It was based upon the fact that the earth's crust is divided into seven massive slabs, or plates, plus several smaller ones. The plates float like islands upon the underneath plastic layer, or mantle. Within the mantle, currents are generated by heat. They start near the superhot center of the earth, then travel upward. Near the earth's surface, they separate and move in opposite directions, just as the currents in a pot of boiling water do.

The people who believed in sea-floor spreading thought that the Mid-Ocean Ridge was located directly above an area where the currents turn and go their separate ways. They thought that those currents pulled sections of the oceanic plate along with them. When the molten material erupts into the rift, they said, it spreads and hardens into new ocean floor. This floor fills the gap left by the traveling sections of plate.

There was one big question to be answered if the sea-floor spreading theory were to be proved—*if* new crust is being formed at the rift, and *if* sections of the earth's crust are moving apart at that point, what happens to the old outer margins of the plates? If they aren't destroyed, the ocean floor and thus the earth would keep getting larger. But the earth *isn't* getting larger, so somewhere the old crust has to disappear. Where does this happen and how?

In 1968, the new oceanic research ship, *Glomar Challenger,* helped to answer that question. This vessel roamed the seas for two years, "reaching back a million years with a coring tube," sending down probes that drilled deep into the ocean floor and bringing up samples of rock. The cores that were taken from near the Mid-Ocean Ridge turned out to be from comparatively new material. The further away from the ridge the samples were taken, the older the material from which they were formed. The oldest cores were taken from the regions that were closest to the continents.

These findings showed that the ocean floor does indeed travel, but the question of what happens to the old crust still had to be answered. It was later found that it is destroyed in the deep slashes, or trenches, that border the Pacific, Atlantic and Indian Oceans. It's here that the plates of the ocean and the plates of the continents collide. It's here that the oceanic crust slides under the continental crust, then disappears into the earth's interior and eventually melts.

THE CORES that were brought up by the *Glomar Challenger* provided many clues about the geological and biological history of our planet. They are formed of layers of material that have been shifting into the sea ever since the earth was created. A study of those layers can reveal events that occurred long before the appearance of man.

One of the earth's long-unsolved mysteries involves the sudden disappearance of the dinosaurs. These mighty creatures existed for millions of years. Suddenly, within a comparatively short period of geological time, they became extinct. Studies of some continental rock layers gave scientists a clue as to what might have happened. The cores that have been taken from the sea floor have reinforced that clue.

The core layers show that the extinction of the dinosaurs happened when there were other great changes occurring on our planet. Thousands of species of oceanic plankton died out, leaving only a few types. Ferns became the predominant form of plant life. The element irridium, which has always been very scarce here on earth, suddenly became plentiful, then grew scarce again.

What caused all of these drastic changes? One theory is that a very large asteroid might have struck the earth. The impact would have caused a cloud of particles to rise into the atmosphere and circle the globe. The cloud would have blocked the sun's heat and caused the temperature to fall, thus disrupting the entire food web.

Another theory is that the cloud could have caused a "greenhouse effect," a trapping of the earth's reflected heat. In this case, the temperature would have risen and there would also have been disruptions in the food web. The impact itself would have caused tidal waves and volcanic eruptions, and these disturbances might have caused even more changes in the weather pattern. During such an upheaval every form of plant and animal life would have been affected. The catastrophe could very well have caused the extinction of the mightiest of all creatures, the dinosaur.

AT ONE TIME, almost everyone believed that the ocean floor was a dead place, an uninteresting graveyard for sunken ships and drowned sailors. We now know that the truth is just the opposite. The sea bed is a place of constant activity and change. It experiences violent volcanic eruptions and earthquakes that send massive swells of water rushing toward our shorelines. It contains fascinating and sometimes grotesque creatures that have adapted in marvelous ways to their dark, cold, pressured environment.

The ocean's waters have provided a cushion between the atmosphere and submarine landforms. This cushion has prevented the erosion and decay that eats into the mountains of our continents. By studying the traveling ocean floor and the raw, eternally new slopes and peaks that rise from it, we can learn much about the history of our planet. We may also gain some ideas about what lies ahead for the human race.

6

The Bountiful World of Inner Space

IN 1979, a group of marine biologists was studying the hydrothermal vents that exist in great numbers along the globe-encircling undersea mountain ranges. These vents spew mineral-laced hot water and seem to be oases in the desert of the ocean floor. When the biologists investigated a vent near the Galapagos Islands, they found an abundance of life clustered around it. Among the variety of creatures, there were some tube worms that reached lengths of up to six feet.

The environment in which these worms live would discourage any previously known form of life. First, the temperatures near the vents may exceed 100 degrees centigrade. Second, there is a great concentration of hydrogen sulfide, which is extremely poisonous.

The biologists wondered how the worms cope with the deadly chemical. To their amazement, they found that these creatures not only tolerate hydrogen sulfide, they actually use the sulfur in place of oxygen in their bodily processes. The discovery was a startling one, because all other known life forms depend either directly or indirectly upon oxygen.

When the Viking lander searched for life on Mars in 1976, its instruments looked only for the elements, gases, and molecules that are associated with life here on earth. When scientists analyzed the data the instruments had gathered, they concluded that there probably is no life on the red planet. Now biologists are having some second thoughts. What if those Viking instruments had been programmed to look for life forms that don't depend on oxygen, those whose bodies rely on other chemical cycles? Would they have found that there is indeed life in outer space?

THE TUBE WORMS of the hydrothermal vents have adapted to their surroundings in their own unique way. The many other creatures that live on the sea floor have done the same. The fact that they survive there is marvelous, because their habitat is the most hostile on earth. While the temperatures near the vents are extremely hot, most of the water at those depths is almost freezing. Pressures amount to as much as seven tons per square inch. The environment is one of eternal night.

Some incredible life forms have evolved in response to these conditions. One species has a horseshoe-shaped mouth that can be used as a shovel to dig into the mud of the ocean bottom. Another has an elongated, rodlike spine that reaches forward to dangle over its mouth. At the end of the spine a lanternlike organ winks and twitches to attract small prey. Some fish look like flowers and are permanently attached to the sea floor. One type of rattail, a fish that has a tapering body, carries millions of luminous bacteria in a gland on its belly. As it creeps along, it looks like a miniature passenger train with many lighted windows.

The benthic animals, those that live permanently on or near the bottom of the sea, have adapted so completely to their environment that they can't exist at a higher level. When some of them are brought to the surface, they react to the decreased pressure by exploding, disintegrating, or twisting into horrible shapes.

The creatures of the sea are experts in camouflage. The ones that live in the night zone are often colored a protective black, black-velvet, or dark brown, so they blend in with their background. Above them, in the twilight zone, prawns, worms, jellyfish, squid, and copepods may be colored various shades to confound their enemies. Copepods, tiny crustaceans that are known as the "insects of the sea," because of their great number and variety, are characterized by a bristled filter chamber behind their mouth.

Some of the fish that live near the daylight, or surface zone, are two-toned. Sharp-eyed predators looking down upon them find it difficult to distinguish their bluish or grayish backs from the water. Predators looking upward at their silvery bellies sometimes can't see them because of the sunshine streaming in from above. That same sunshine may also cause surface fish to appear as dark objects to the creatures in the lower water layers. Predators attacking these objects may find only a piece of wood, but they may also find a tasty meal.

LIFE IN its various forms is continuous from the top layers of the ocean to the sea floor. The amount of life at every level depends on how much vegetation is growing at the surface. At sea, as on land, the animal kingdom depends upon the plant kingdom for food.

A large part of the ocean's plant and animal life consists of microscopic organisms called plankton. Since plankton need the sun's light and energy, they float near the surface to depths of only about three hundred feet (one hundred meters) . Despite their tiny size, plankton are the basis of the ocean's food web, and thus have been called the world's most important nutrient.

Just as the land plants depend upon the minerals in the soil for their growth, plankton depend upon the minerals in the sea. During the winter, strong winds stir up the water, and a rich supply of mineral-laden bottom water is brought to the surface. As the winds continue to blow, the minerals become thoroughly mixed throughout the ocean's uppermost layer.

With the arrival of spring, and its increasing warmth and daylight hours, single-celled algae multiply rapidly. Within days the sea is tinged a rich brown, yellow, or green. The animals in the region gorge themselves on these living carpets.

By summer the storms have ceased, the upwellings have died down, and the top water layer has become warm. This warm water floats on the denser, colder water beneath it. As the two layers become more stabilized, a pronounced border forms between them. The border is known as the "thermocline," and marks the depth at which there is a sharp drop in temperature and a related sharp rise in density. The thermocline forms an impenetrable barrier that prevents any further mixing of the warm light surface water with the cold, dense water beneath it. Below the thermocline, the water gradually becomes colder and denser. The coldest, densest water is found at the bottom of the sea.

When the thermocline is established, no more mineral nutrients can enter the top water layer. The remaining algae are consumed by animals, and by the end of the summer the food supply is scarce. In autumn, the water is again stirred by storms. The thermocline is pierced and upwellings bring minerals to the surface. There's still plenty of sunlight, so the plants reproduce and another "bloom" occurs. Then, as the sun sinks lower in the sky, the days grow shorter, and the number of plants again decreases. Soon, only the resting, or dormant spores are left. Until spring returns, they drift in a sort of suspended animation.

EVEN DURING the height of the growing season, life isn't smoothly distributed throughout the world ocean. In the central Pacific, there's a desert that spreads over much of the central gyre. The water there is clear and blue, because there are few plant and animal organisms to give it any other color. The pastures of the sea—the places where ocean crops flourish and where fish and mammals congregate—are located where offshore winds move the water, and where strong currents converge to form upwellings.

The cluster of animals in these highly populated places can cause a problem for sonar operators. Their sea floor soundings may be confused by "ghost echoes" from shallow depths of from 500 to 1,500 feet (175 to 500 meters). These "phantom bottoms" are caused by the "deep scattering layer," a congregation of various types of sea creatures that move up and down in response to the presence and absence of sunlight. During the night, it's positioned close to the surface. As dawn breaks, it moves downward.

IN ITS VARIETY and interest, the life of the sea equals and often surpasses life on land. For instance, the remarkable annual migrations of some birds and animals is overshadowed by the migration of a form of the lowly eel. This creature leaves the rivers of Europe, travels across the Atlantic Ocean to a certain section of the Sargasso Sea, then spawns and dies. The eggs are caught up in the Gulf Stream, and three years later arrive at the European coast. Here they hatch, and the infants swim upriver to spend most of their adult lives. Then, at a signal from their biological clocks, they head back to the Sargasso Sea to start the cycle again.

No one can guess at the total number of individual animals that live in the sea. Scientists haven't even finished their cataloguing of the different *kinds* of creatures that exist there, because new ones are still being found. We do know that oceanic life ranges from the microscopic plankton to the Antarctic blue whale, which is much heavier than any dinosaur that ever lived. We know that life in the sea ranges from the most grotesque forms to the terraces of glowing coral that spread along tropical coasts. We have seen that the sea harbors some forms of life that evolution seems to have passed by. Jellyfish, sponges, starfish, and horseshoe crabs are all living fossils that still flourish in the world ocean.

It's certain that our study of the sea will turn up many more surprises. The discovery of the strange tube worms that live in the toxic atmosphere of the hydrothermal vents was only one of the more recent surprises. Those creatures may have provided us with a clue in our continuing search for life in outer space. Is it possible that part of that search will take place in the many-faceted world of our own sea's inner space?

7

"The Winds Blow and the Waves Come"

THE CAPTAIN of the Canadian steamer *Princess Sophia* was uneasy as his vessel crept along Alaska's Lynn Canal. It was three o'clock on an October morning in 1918, and the combination of darkness and fog made it impossible to see where he was heading.

A short while later, the captain's sense of foreboding was justified. The ship, a full half-mile off course, struck a reef. An inspection of the damage showed that the hull of the *Princess Sophia* had been penetrated by a jutting point of rock. Water was seeping into the hold, but at a very slow rate. There would be no reason to panic as long as the pumps continued to work and the sea remained calm. The captain radioed for assistance and readied the crew to meet the emergency.

When a storm broke out a few hours later, the pumps were still performing, and the ship remained firmly anchored to the reef. Nevertheless, the captain's next radio message stressed the urgency of the situation.

The storm was at its peak when another ship drew close enough for its lookout to spot the *Princess Sophia* through a cloud of gray ocean spray. There could be no rescue attempt until the wind died down. Still, the embattled vessel was securely impaled upon the rock and its passengers and crew faced no immediate danger.

A short while later, however, before the horrified eyes of the rescue ship's crew, the *Princess Sophia* was suddenly lifted off her rocky anchorage. A flood poured through the gash in her side, and within minutes the steamer floundered and sank. Three hundred and forty-nine people died that day, but not because of the battering of the waves. The cause of the tragedy was the gentle, persistent, and wholly predictable action of the tides.

THE ACTION of waves and tides has always had a great effect on the lives of humans. Long before their movement was fully understood, they were looked upon with awe and sometimes fear. Many ancient observers believed that the tides were caused by a serpent, a monster so enormous that it encircled the globe. Twice a day, about every twelve hours,

> *Beneath the lashing of its tail,*
> *seas, mountain high, swelled upon the land.*

Later, the Greek astronomers reasoned that there must be a connection between the ebb and flow of the tides and the appearance and disappearance of the moon. When Isaac Newton developed his theory of gravity, the Greeks were proven right. The moon's gravitational pull causes an increase in the sea level, not only on the side of the earth facing the moon, but also on the opposite side. Because of the earth's rotation, most places on earth experience two tidal cycles every twenty-four hours.

The sun also exerts a gravitational pull on the sea. However, because it's much further away than the moon, its effect is considerably less. When the sun and the moon are in a direct line, the sun's pull reinforces that of the moon. At that time, unusually high, or spring, tides occur. When the sun and the moon work against each other, low, or neap, tides result. The earth's rotation, the inclination of the earth's axis, and the alignment of the planets also have an effect on tides.

On the open sea, the coming and going of the tides is barely noticeable. The same is true of the islands and sections of continents that lie near the equator. On the long, gentle sloping beaches of the South Pacific islands, the tidal range may be only a few inches. Here there's usually little to fear from the incoming water, and homes are built close to the shoreline. Occasionally, however, tidal waves will destroy those buildings. The combination of a high tide and a violent storm always holds the potential for disaster.

There are many coastal areas where the ordinary everyday tides are spectacular and dangerous. The incoming tide can be blocked by cliffs that form a narrow opening to a bay or an estuary. As the water pours through the gap, a "tidal bore" is formed. In one Chinese coastal region, the tidal bore takes the form of a fifteen-foot wall of water that moves upriver at fifteen miles an hour. In Germany, the city of Cologne is located 120 miles inland. Despite its unlikely location, it's a busy seaport because the tidal action permits ocean-going vessels to navigate the Rhine River twice every day.

The tides of Nova Scotia's Bay of Fundy are perhaps the most famous in the world. When the tide comes in, water from the Labrador Current, the Gulf Stream, and another strong current flows into the bay. At the bay's northern end, the water converges on a restricted channel. As the tidal bore moves up the channel, in some locations the water level rises an astounding forty feet within a few minutes. At one point the force of the incoming water causes a waterfall to reverse its direction. At high tide anchored fishermen's boats bob on the surface of the water, while low tide leaves them stranded on the mud of the channel bottom. The people who live and work near the Bay of Fundy plan their daily activities around the movement of the tidal waters.

Fortunately, tidal action is so dependable that it can be charted far in advance. All over the world, clam diggers and fishermen consult tide charts. Marine navigators plan their trips so the high tide will float their vessels safely over reefs and shoals. People in the ocean salvage business use the buoying force of high tides to help lift and float sunken ships.

In addition to having an effect on the day-to-day lives of human beings, tidal activity has determined the outcome of crucial moments in history. Part of the Allied success in the World War II coastal invasion of France was attributed to a careful study of the tide charts. Julius Caesar was not as well informed. When he planned his first invasion of Britain, he was familiar only with the low tidal ranges of the Mediterranean and Adriatic Seas. As a result, his men failed to moor their boats far enough up on the English shore and when the high tides came, Caesar lost a big part of his fleet. He had to withdraw his forces and return the following year to continue the battle.

Marine animals are also affected by the tides. With the first hint of seaspray from the incoming water, mussels and barnacles extend their cilia, tiny hairs that direct food into their mouths, in anticipation of a meal. Anemones cling to rocks, so they won't be carried inland. Certain worms burrow underground to escape the abrasive force of the sand-laden water. Others dig themselves out so they can feed upon the sea's nutrients. Shore birds follow the edge of the incoming water to dine upon the life that seems to spring from the surf. Snipes run back and forth along the water line in frantic bursts of energy. They are looking for the little puffs of air that are expelled by tiny creatures that live in the sand.

As the tide recedes, the barnacles and mussels retract their feelers. The birds follow the receding water's edge. The inhabitants of the tide pools settle down to await the next tide and the food it supplies. The fish and worms that live in the depths of the sea will soon be feeding on the nutrients that the ebbing tide has extracted from the land.

THE IMMENSITY of a tide is hard to envision because it encompasses an entire ocean. In contrast a wave moves in only a small part of the sea. Nevertheless, these swelling, crashing, foaming "white horses of the sea" have captured man's imagination since the first time he saw them.

Most waves are caused by winds pushing upon the surface of the water. Gentle passing breezes form momentary ripples. Stronger gusts form choppy whitecaps. Hurricanes and typhoons form destructive waves that may range from twenty to forty feet in height. Towering, savage waves that can sink large ships result from gale velocity winds that have traveled over a vast "fetch," at least 200 to 700 miles of open sea. The waves that result from localized storms don't have enough fetch to build up to such a great height.

The wave's journey ends when it reaches a shoreline. Here the ocean bottom rises, producing friction, or drag, in the lower portion of the wave. This lower portion slows, while the upper part of the wave continues to rush forward. Close to shore, the top of the wave steepens, crests, and curls forward, finally breaking into a crashing, tumbling white foam blanket that spreads out over the beach.

Sometimes their journey's end is more dramatic. If a wave slams into a rocky coast, and the wind and the curve of the sea bottom are just right, the water will leap upward to a height twice as great as that of the tallest waves on the ocean itself.

At other times, a breaking crest may curl over rapidly and trap a pocket of air. As the curl tightens, the air is compressed. As it breaks loose with a roar, it sends up a great geyser of foaming water.

When the surf is high, seaweed is tossed about and furrows are worn into the sand. Salt spray, the "dust of the sea," fills the air, while beneath the water's surface, sand swirls and the seabed ripples. Surfers, who catch the motion of the swell several moments before it peaks, swoop and glide as they move with the circular motion of the water inside the wave form. They actually become a part of the power that lies in the sea.

The same power that can be so enjoyable to a surfer can also be extremely destructive. In 1937, hurricane waves in the Bay of Bengal destroyed twenty thousand boats and took over three hundred thousand human lives.

MANY A SAILOR has bragged about riding his vessel through a hundred-foot wave. Until recently, most such reports were thought to be exaggerations, because ships rarely encounter waves that are even thirty or forty feet high. The problem with measuring a wave is that when someone is standing on the deck of a ship that's nosing downward into a high sea, it's likely that the wave will look twice as high as it really is.

Occasionally, there is an accurate yardstick by which the wave can be measured. In 1933, a United States Naval vessel was caught in a Pacific gale. As the ship was lifted by a wave, a man on the bridge saw that its crest lined up almost perfectly with the crow's nest on the mainpost. From the trough, which is the open space between waves, to the crest, the wave measured an astounding 112 feet, the height of an eleven-story building. Fortunately, the mountain of water picked the ship up and passed under it. If the wave had crashed down upon the vessel, no one would have escaped to tell of its height.

Such an instance does indicate that there really are

such things as "superwaves." Scientists using computers and laboratory wave-making equipment have shown how a monstrous tower of water can come into existence. When tides, and winds, and currents all interact in just the right way, the wave that is formed may be twice as tall as any ordinary storm-created wave. The circumstances that result in a superwave are unusual and complex and, fortunately, don't occur very often.

It's now believed that such "killer waves" may have acounted for the disappearance of many large, almost unsinkable ships. During World War II, when the *Queen Elizabeth* was being used as a troopship, it was hit by what is now thought to have been a killer wave. That huge passenger liner rolled to within a degree and a half of its "critical point." If it had been pushed just a little further, nothing could have kept it from capsizing.

Today, satellites are being used to study possible breeding grounds for killer waves. So far, none have been found in the Bermuda Triangle, but some "hotbeds of dramatic wave activity" have caused many ships to founder off the coast of Norway. Here high winds occasionally combine with unusual current conditions to form killer waves. There's also a "demon sea" near Japan. Recently, in that region, a thirty-foot chunk of metal was torn off the bow of an ore freighter by an unusually high wave. In 1969, the same type of wave activity broke a thirty-four-thousand-ton ship in half.

The Japanese think that their killer waves are

caused by cold Siberian winds that skim over the relatively warm waters of the swift Kuroshio Current. This action, combined with deep and unpredictable currents, seem to be able to stir up boiling cauldrons of deadly high seas.

Japanese scientists are now overseeing the construction of a wave observation robot. When it's put into place, it will be able to measure movements in the sea to depths of 16,400 feet. Perhaps with the help of the robot, oceanographers will be able to predict some of the more dangerous wave activity, and lives will be saved. Meanwhile, the men who sail the Demon Sea will have to depend on the patrol ships and helicopters that are standing by for emergency calls.

A TSUNAMI is an extremely dangerous wave that is caused by an undersea earthquake or volcanic disturbance that sets a great volume of water into motion. This large, solitary wave can cover long distances at tremendously high speeds. When a tsunami hits a shoreline and releases its energy, it can easily take hundreds of lives and destroy millions of dollars worth of property.

The most destructive recorded tsunami occurred in 1883, when the volcanoes on the Indonesian island of Krakatoa had a series of powerful eruptions. During the third explosion, a large part of the island caved in. After the fourth one, a tsunami rushed away from the area at nearly 350 miles an hour. As it swept across nearby islands, it took thirty-five thousand lives.

To protect people from tsunamis, observation posts have been set up to track them as they move across the sea. The Japanese have become the foremost experts in tsunami watching and prediction, because their land is often subjected to the fury of these lethal waves.

Although it's hard to believe, the tallest waves of all don't occur on the surface of the sea, but beneath it. Internal waves are found along the boundaries between water layers of different densities. The waves along the thermocline can rise to heights of two to three hundred feet and are capable of traveling across an entire ocean basin.

Scientists don't know the exact forces that set internal waves into motion. It's thought that some of them are caused by weather disturbances. Perhaps others start when a ship plows through the water. It is known that these unseen waves have an effect on the undersea transmission of sound, and they probably provide much of the energy for stirring up the water above the thermocline.

WE ARE GOING to have to have to learn much more about internal waves as we spend more time exploring and working in the sea. Meanwhile, it's the surface waves that most affect our lives. They are, after all, the ocean's troops in the ongoing war between the sea and the land.

When they break upon the shore, the beach itself becomes a sort of constantly moving "earth river." Sand and silt are moved about. Sandbars, reefs, and lagoons are formed. The beach is widened at one point, while a few miles away, it's eroded. In one experiment, the sand in a certain region was soaked with red dye. Within a few days, some of the colored sand was found one hundred miles away.

A United States Army brochure once stated, "Our campaign against the encroachment of the sea must be waged with the same care that we would take against any enemy threatening our boundaries." This ambitious statement didn't change the fact that man's efforts to control the effect of tides and waves are still largely futile. Seawalls, jetties, and breakwaters eventually fall. One Scottish breakwater was capped with huge stone blocks, cemented, then tied together with heavy iron rods. The resulting structure weighed 1,350 tons, but the power of the water eventually carried it onto the shore that it was supposed to protect.

We've had no better luck with our lighthouses, harbors, piers, and coastal roads. Storm-driven waves have tossed boulders through the roofs and windows of buildings that were over a hundred feet above the usual water level.

Since there's not much point in fighting the power of the sea, we might as well learn to work with it instead of against it. Until we can harness and direct that power, we'll have to listen to the simple words of the old seafarer.

"The winds blow, and the waves come," he said. He knew from long experience that there's not much we can do about that state of affairs.

8

The Sea Is a Living Soup

MILLIONS OF people all over the world suffer from malnutrition. Many parts of the world experience frequent droughts, which often lead to deaths from starvation. One solution to both lack of food and lack of water may lie in the sea.

A few years ago several oceanographers thought of a fanciful way to supply water to desert areas. Since Antarctic icebergs are a part of the continental icecap, they are free of salt. Why not tow them to a region that needs water, such as Saudi Arabia, and ground them on offshore shoals? As the icebergs melt, the runoff can be piped into reservoirs, then used to irrigate fields.

Most of the people who heard about this scheme

laughed and called it an impossibility. A handful of more imaginative scientists took a closer look at it. They planned an imaginary trip from Antarctica to southern California, using three ocean-going tugs to haul an average-sized iceberg. After following the Humboldt Current, they used a roundabout route to take advantage of other strong currents. The journey would take about a year, and during that time the iceberg would melt down to about half of its original size. Even so, it would still be able to furnish over 300 billion gallons of fresh water. In an area that experiences frequent droughts and is sometimes forced to ration water, "Operation Iceberg" sounds like a good idea.

Good idea or not, many decades will probably pass before there will be any icebergs glinting under the California sun. Meanwhile, the search for other ways to obtain fresh water from salty, mineral-laden ocean water continues. At least a dozen workable methods are already known, but most of them cost too much to be used in anything but an emergency situation. Some experts have tried to freeze ocean water to obtain fresh water, but so far there's no practical way to separate out the pockets of salt that remain. One promising method uses a membrane tube that permits water, but not salt, to pass through. Some scientists have suggested tapping the freshwater springs that spurt from the ocean floor.

Key West, Florida, is a coastal city that has a large

population and little natural fresh water. Like hundreds of similar places throughout the world, it has a desalinization plant. The fresh water the plant produces from seawater isn't cheap, because the processing requires a lot of expensive energy. It also requires complicated, corrosion-proof equipment.

Mexico has a desalinization plant that runs on solar energy, so its operation is more economical. Nevertheless, the converted water still costs more than fresh water from natural sources. The Israelis, who are leaders in the search for ways to use water from the sea, don't consider the cost of their converted water as a burden. In their desert region, they are happy to have water at any price.

As the world's population increases and the need for water becomes more urgent, we'll have to find economical ways to process sea water. When we do, the threat of water shortages will disappear, because ninety-seven percent of our planet's water supply lies in the ocean. By drawing upon that abundance, we can be assured that no crop will ever shrivel and die because the rains didn't come. Water will never again have to be doled out. Every desert region could be turned into an orchard, a pasture, or a field of grain. Finding a solution to the water shortage will give us a means of solving the world's food shortage.

A FASTER SOLUTION to the food shortage lies within the

ocean itself. Its untold millions of tons of marine life outweighs all of the life that's to be found on land. Within its waters we can find enough protein to supply a human population several times larger than the one that already exists. The ocean can also supply a large part of our vitamin and mineral needs.

To fully tap this abundant food supply, we're going to have to make a break with the past. Long ago, people found that it was easier to grow their own grain and to raise their own cattle than to roam across plains and through forests looking for the next meal. One of the marks of advancing civilization was the change from hunting to farming.

Today's fisherman uses sonar and aerial reconnaissance to locate schools of fish. He studies fish migration patterns and knows about the tricks that currents can play. On his ship he has a freezer to keep his catch from decomposing. Some fleets are accompanied by a factory ship, upon which the fish are filleted, canned, and frozen and their by-products processed at sea.

All of these modern techniques don't change the fact that most fishermen are still hunters, not farmers. If the temperature or the salinity of the water in a fishing region changes, or if for some reason the supply of nutrients disappears, the fish leave the area and congregate somewhere else. And it's not easy to find a school of fish in the vastness of the sea.

By hunting their prey, fishermen are now catching fifty or sixty million tons of fish every year. Some experts say that if they became farmers, they could get twice that amount.

"Aquaculture" includes not only the growing of fish, but also of aquatic plants. There's a lot of nutritious eating in some types of seaweed. Puddings made from a species of kelp, rye flour, and milk saved the lives of many babies during the Irish famine. The Japanese start off many of their dinners with seaweed soup.

The Japanese have practiced aquaculture for centuries. They grow seaweed on bamboo poles and also anchor it to rocks in the bottoms of shallow pools. They are pioneers in the art of growing seafood in their coastal waters. In addition to their oyster farms, they cultivate other shellfish, as well as sea urchins, anemones, and various kinds of fish.

Off the coast of southeast Asia, farmers have constructed floodgates, which allow fish to ride toward shore on the incoming tide, but trap them when the tide recedes. Like cattle on a range, the fish feed upon the nutrients that appear naturally in the water. If the food becomes scarce, the farmer supplements it.

Fish don't necessarily have to be fenced in to be farmed because they will normally stay where the conditions are right and where they can get enough to eat. Nevertheless, most farmers put nets across the mouth of the bay or estuary to keep predators out and to keep the fish in a well-defined area. The barriers also allow the farmers to limit their stock to one or two desired species.

When the fish are kept in a confined area, the farmer has to harvest his "crop" regularly, or the stock will become overcrowded. Fish like space around them, just as other animals do. If they can't move freely, they may stop eating, refuse to breed, and sometimes die.

Getting fish through their early growing stages is a tricky job. In a fish hatchery, the temperature and chemical composition of the water must be kept at a constant level, and the babies must be fed the right amount of specially cultured food at the right time. To meet those strict conditions, the young fish are kept in a tank and constantly monitored. A well-run fish hatchery produces many more healthy adult fish than does the nursery of the open sea, where predators lurk and where famines can occur.

RAISING FISH isn't a job for amateurs or for people who are easily discouraged. One group of people started a fish farm on the west coast of Scotland. After fencing off the bay and installing sluice gates, they started to construct a concrete wall on one side of the farm. When the project was near completion, several marine biologists went to a nearby fish hatchery to pick up a half million infant plaice, a flavorful flat fish. On the return voyage, the biologists spent most of their time dashing around with oxygen cylinders, blowing the life-giving gas into the plastic bags that contained the fish.

Meanwhile, the people who were working on the wall were having some problems, so the work wasn't done when the fish arrived. Dry cement was still being washed into the bay from the burst paper bags that littered the shore. There wasn't time to clean up the water because the fish would die if they stayed in the plastic bags any longer.

Surprisingly, the baby plaice survived the shock of their drastically changed environment. They adjusted to the fluctuations in temperature, the changed salinity, and the cement-laden water. Unfortunately, some of them didn't survive the attacks of the crabs that lay in the mud and sand at the bottom of their new home. The farmers had to plow up three acres of ocean bottom and resand it to get rid of the predators.

It had been a rough start for the new business, but the farmers remained optimistic until the rains started to fall. After days of the downpour, the low moorland that bordered the bay began to drain into the sea. The runoff pipes that were supposed to channel the rainwater outside of the enclosed farm area became overloaded. When the bay was flooded with fresh, peaty water, the acidity and oxygen content of the fish farm began to fluctuate wildly. The farmers began to think that they were doomed to failure.

When the rains finally stopped, they felt a little better. Amazingly, most of the fish were still swimming around. The farmers still had one more unpleasant surprise facing them though. They soon learned that fish apparently have a pecking order, just as chickens do. When they were fed, the ones at the top of the order gorged themselves, while the ones at the bottom of the order had to make do with whatever was left. This last discovery undermined the farmers' main goal, which had been to furnish the market with a steady supply of uniform-sized fish.

Despite all of the problems, this fish farm flourished, and eels and other sorts of fish have been added to the stock. The farmers' perseverance and hard work have paid off.

FINDING A suitable place for a fish farm isn't as easy as it may seem, because coastal waters are used for so many of man's other activities. Also, much of that water has become too polluted to be used for raising fish. Surprisingly, some forms of pollution may turn out to be blessings in disguise. Some fish thrive on the nutrients in the garbage that's dumped into the sea.

Scientists aren't sure about the effect that the thermal pollution from power plants has on marine life. In many cases, the raised temperature of the water has caused the fish in the region to mature faster, grow larger, and become ready for market sooner than they ordinarily do. The skin divers who explore the shallow water near some nuclear power plants are delighted by the number and variety of sea creatures and plants that they find there. No one knows what effect this plentiful growth will eventually have on the marine life in adjoining areas. Long-term studies must be made before the final verdict about thermal pollution can be made.

THE OCEAN has been called a "living soup," teeming with life, stocked with an abundance of food for both human beings and animals. So far, we have barely begun to take advantage of what the sea has to offer. Before we can, we may have to learn to eat unfamiliar plants and strange fish. Marine farmers will have to deal with the same sorts of discouragement, frustration, and hard work the land farmers do. The problems of transporting the harvest of the sea to the people who need it will have to be solved.

As land becomes scarce, aquaculture may become as common and as accepted as agriculture. A hungry world can turn to the sea, because, properly managed, the world ocean is a food basket that still refills quickly.

9

The Last Great Gold Rush

In 1921, the government of Germany was faced with an overwhelming financial problem. The Allied nations, which had defeated Germany in the recently ended World War I, presented the country with a bill for the damages Germany had caused by starting the war. The reparations added up to a grand total of 320 million gold marks. With their industry and economy in ruins, German leaders could see no way to come up with that amount of money.

By 1924, it appeared that there might be a solution to their problem after all.

"There's an abundance of gold in the ocean," said Nobel Prize winner Fritz Haber. "If we can find a way to extract it, we would have not only enough money to pay the reparations, but also enough to rebuild our country."

With enthusiasm and high hopes, the Germans collected enough money to dispatch a research ship. Its job was to survey the mining fields of the Atlantic and to find an economical and efficient way to reap their rich harvest. Two years and millions of dollars later, Haber had to admit failure. He had overestimated the amount of gold in the waters of the ocean. He had also underestimated the amount of money it would take to recover that gold. He had been no more successful in his quest than the ancient alchemists had been in their efforts to turn base metals into precious ones.

FRITZ HABER had actually been right about one thing—there *is* an abundance of gold in the ocean. There's also an abundance of every other metal known to man, because the rivers of the continents are constantly dumping large amounts of minerals into the sea. Submarine volcanic eruptions add still more, as do the winds. All types of dissolved minerals are accumulated faster than we could possibly use them.

The problem, as Haber discovered, is that while the world ocean contains untold millions of tons of minerals, it contains many times more tons of water. In this case, the apparently rich "soup" is much more like a very lean broth. To obtain only a few ounces of gold, thousands of pounds of water would have to be processed. Such an operation would require tremendous amounts of power in a world that's already experiencing a severe energy shortage.

There have been a few substances that have been economically retrieved from seawater. For five thousand years men have been collecting tidal water, letting it evaporate, then collecting the salt that remains. In recent years, scientists have found inexpensive ways to extract magnesium, bromine, and potassium. They have been studying certain sea creatures that are experts in the retrieval of selected minerals from the water that surrounds them. The lobster, for instance, concentrates copper in its body, while there's a type of sea worm that absorbs vanadium.

Someday, perhaps, we'll be using an advanced technology to obtain gold and other precious metals from the sea. Until then, it appears that their extraction will cost much more than we can afford to pay.

MEANWHILE, we're having better luck in mining the buried wealth of our continental shelves. For hundreds of years, with only the most primitive of tools, people have been collecting shell, sand, and gravel from the shallow coastal sea floors. From submerged river channels, they later mined tin and platinum and dredged nodules of phosphorite.

The Japanese are currently getting about thirty

percent of their coal supply from shafts that begin on dry land, then extend far under the continental shelf. The seaward ends of some of the shafts are thousands of feet below sea level.

It was during the 1950s that the appearance of offshore oil rigs signaled the beginning of one of the greatest treasure hunts in history. The continental shelves contain a large part of the world's current oil reserve. Most of it hasn't yet been found, but as the supply of oil on land diminishes, more and more drilling will be done along the world's coasts. In the United States, the demand for oil rigs has become so great that our shipyards can't keep up with it. Many of them are now being built in Japan and other countries, then towed to the drilling sites.

Since most of the drilling is still being done in water that's 600 feet or less in depth, the oil production platforms are usually anchored to the sea floor with piles. As the search for oil extends further out to sea, floating platforms will become more common.

Further in the future, there may be unmanned deep-submersible robots that can work around the clock. One robot has been designed to "feel" its way along a length of pipe until it reaches the end. It then picks up another piece of pipe and welds the two lengths together. Another robot can lay pipe within a trench. In one continuous operation, it shoots a stream of water from a huge fire hose nozzle to scoop out the trench, lays the pipe, then covers it up with sand. Many of these undersea robots are equipped with television cameras and interchangable "hands," with which they can drill holes, cut cables, and install parts.

Robots can be a big help, but they will never take the place of human beings. It's men and women who will have to cope with the emergencies and dangers that accompany life on an offshore drilling platform. Even when the sea is calm, it's always on the move, and the constant tug of tide and current can damage a platform. The perils increase a hundredfold when the weather is bad. Violent storms occur in the Arabian Gulf. Ice floes create serious problems off the coast of Alaska. Hurricanes wreak havoc in the Gulf of Mexico. Rigs are battered and torn by waves, and workmen are injured.

Sometimes the storms are so severe that an oil rig collapses. In 1979, eight men died in the Gulf of Mexico because of such an accident. Just one year later, a sudden gale blew up twenty-five-foot seas in Norway's North Sea oil fields. The supports of one rig couldn't stand the strain.

"I heard a crash like thunder," a survivor later said. "The rig shook. I jumped into a lifeboat, and as we lowered it, the whole rig toppled over."

One hundred and twenty-three men died in that disaster. In 1982, eighty-four more died when the *Ranger*, one of the world's largest production platforms, listed, then sank in a howling storm off Newfoundland. Investigators later found the rig lying on its side in 300 feet of water. One of its supports had snapped in two.

With each succeeding tragedy, we are reminded of the danger that awaits human beings as they venture away from shore. We realize the high price the sea extracts for giving up its hidden treasures.

THE SEARCH for oil may one day be rivaled by a combination of other mining possibilities. During the 1960s, some oceanographers were probing Africa's Red Sea. This body of water is located over a continental rift, which is much like the rift valleys of the ocean. As the scientists took a series of temperature readings, they noticed something unusual. Despite the fact that hot water usually rises to the surface, the thermometers showed that there were three locations near the sea floor where the temperature rose dramatically.

Further observations showed that the water in those locations was being heated by periodic volcanolike eruptions from the rift. It didn't rise because, despite its heat, it remained very dense and heavy. The oceanographers found that high density was caused by a great concentration of zinc, copper, gold, and silver. The sediments under one of these "brine pools" was like black tar and proved to be too hot to touch.

Plans are being made to pump these brine pools to the surface so their minerals can be retrieved. Meanwhile, the search continues for other such rift areas. During one exploration, the submersible research vessel *Alvin* was off the coast of Ecuador. The oceanographers onboard were amazed to see several volcanic vents spewing forth mineral-laden hot water. A nearby sea floor copper deposit was ten times richer than any deposit ever found on land. But even with that great mineral concentration, the costs of undersea mining today are too great to consider retrieving it. Nevertheless, the search for more such volcanic vents is continuing. Perhaps some of them *will* have enough minerals to make a mining project worthwhile now or in the future.

BRINE POOLS and volcanic vents may eventually give us some of the minerals we need, but the most exciting current possibility are the minerals that have been found on the abyssal plains. The muds and clays that cover a large part of the Atlantic and Pacific sea floors are rich in iron, copper, nickel, and cobalt. More easily obtained and even more valuable, however, are the strange, charred-looking rocks, or nodules, that were brought up and largely ignored by the crew of the HMS *Challenger*. They have recently caused engineers to make breathtaking advances in deep-sea technology. They have caused heated arguments in many international Law of the Sea conferences. They may be the focal point of the last great "gold rush" to take place on our planet, because these nodules contain enough magnesium, cobalt, and nickel to supply the world's needs for hundreds of years.

In some ways, the nodules are a puzzle to scientists. Why haven't they been buried under the constant fallout of material that descends to the ocean floor? How were they formed? How widespread are they?

It appears that the materials from which the nodules were formed came from sediments that were carried into the sea by continental rivers. Perhaps some of the sediment clusters around a nucleus such as a shark's tooth, a hard bit of clay, or a tiny fish bone. As millions of years passed, layer after layer of mineral material may have been added.

The average nodules range from the size of a marble to the size of a small potato. Some of them are much larger. In 1951, a crew from the Scripp's Institute of Oceanography brought up a giant-sized nodule that was almost three feet in diameter. The men were ecstatic over their discovery.

"It is the sort of specimen that should be placed on a pedestal in a museum somewhere," one of them said. "We carried it into the lab, grinning like Cheshire cats."

WHILE THE potential rewards of deep ocean mining are great, so are the problems. Before any work can begin, the area to be mined must be thoroughly mapped and many nodules brought to the surface and tested. The operation is expensive and difficult, so only the richest sites can be selected.

Just building the equipment to *install* a sea floor mining device takes a lot of planning and money. Special anticorrosive materials must be used. New technology must be developed. The operation of the machinery must be tested again and again for dependability.

The actual installation will also be hazardous and tricky. Every step will have to be done by remote control, using television cameras and robots. Expensive equipment may be lost, with little chance of recovery. Sea floor currents may tangle lines or make them drift far from their planned location. While monitoring the operation, the ship's crew may have to battle stormy weather and struggle to keep their vessel on station.

The same problems that may plague the installation will continue to plague the actual mining, which will probably be done by a huge caterpillar tractor. It will be guided by television cameras and attached to the ship by a cable. Controlling such a long cable won't be an easy job, especially in strong winds and high waves. Although the mining device itself must be sturdy enough to withstand the conditions of the deep sea, it must be light enough not to get bogged down in the soft mud of the ocean floor.

This device will possibly run back and forth into the current, making swaths just as a farmer plows a field. As it moves, a large vacuum attachment will inhale the materials from the sea floor. The force of rushing water will push the nodules through a tube to the surface ship. The unwanted sediment will be blown out of the back of the machine, leading to another possible problem. Scientists don't know how long this cloud of sediment will remain suspended, or what effect it will have on the creatures of the sea floor. It might cause a disruption in the food web or bury a variety of small creatures. These ecological aspects must be thoroughly examined.

The people who are exploring the possibilities of mining the ocean floor are aware of the dangers and the problems. They are also aware of the great profits that could be made. One United States company has already applied for a claim in a section of the central Pacific. Our government didn't issue the permit, because no one knew if such a project would be allowed under the existing international Law of the Sea.

The uncertainties aren't stopping prospectors from continuing to explore the possibility of doing this type of work. Recently, one of them took his "old bucket" of a mining ship on a shakedown cruise. As the vessel left port, he was brimming over with enthusiasm.

"Back in the 1880s a lot of folks left St. Louis," he said. "To them, what they were doing was all trial, all new. That's the way it is with industrial oceanography today. Deep-ocean mining is the last great gold rush.

"Ten years ago, people laughed at the idea of bringing up minerals from fifteen thousand feet deep. Well, we've already proved that it can be done. The other miners have stopped laughing a long time ago."

THIS NEW BREED of miner already knows the advantages of his developing profession. He'll have no rocks to blast, no shafts to sink, no town sites to be constructed for his workers, no cave-ins. The exact quantity of ore in any one region may easily be measured by just scanning the seabed with a new sonar device. As Arthur C. Clarke said, "Here we go grubbing about in the earth for our metals and chemicals while every element that exists can be found in the sea. The ocean is, in fact, a kind of universal mine that can never be exhausted."

Many people are offering words of caution to counterbalance such enthusiasm. They point out the great difficulties that are unique to ocean mining. They say that there are still many questions that haven't been answered. Said one oceanographer: "All we really know now is that our ignorance of this resource is vast."

"We must decide to protect and manage this resource," said another. "Once we have begun mining, we are committed. We have bitten into a million years of history for the first time. If this is, indeed, the last great gold rush, let us apply to it the lessons we've learned from the mistakes we've made on land. Let's not plunder the sea as we have plundered the land."

Members of the Coast Guard fight an oil rig fire off the coast of Louisiana. U.S. Coast Guard

These Pacific Ocean manganese samples are about four inches in diameter. Woods Hole Oceanographic Institution

The broken halves of a tanker swirl in a sea of foam. The ship had been carrying a cargo of heavy fuel when it ran aground off Nantucket Island. The crew was airlifted to safety, but the accident resulted in a major oil spill. U.S. Coast Guard

This manganese nodule was magnified 600 times to show the broken domes that were built by marine organisms. The nodules may owe their existence to such organisms. Scripps Institution of Oceanography

Many beaches have been temporarily despoiled by the remains of oil slicks. U.S. Coast Guard

Workers clean the last of the tarballs from the beaches of Sandy Hook, New Jersey. The tarballs were the result of an oil spill when a barge and a tanker collided. U.S. Coast Guard

This chamber is being lowered to a shallow sea floor, where it can be left for up to a month. This experiment will provide information on how bottom-dwelling organisms contribute to the movement of pollutants. Woods Hole Oceanographic Institution

WOODS HOLE
OCEANOGRAPHIC
INSTITUTION

ALVIN

Alvin's *crew must learn to get along in cramped quarters. Woods Hole Oceanographic Institution*

The deep submersible research vessel Alvin *is shown here with its mother ship,* Lulu. *Woods Hole Oceanographic Institution*

The Alvin's *searchlights illuminate the inky blackness of the underwater world. Woods Hole Oceanographic Institution*

Scientists and members of the ship's crew help recover a float used in ocean engineering studies. Woods Hole Oceanographic Institution

Seasat mappings of the world ocean show how heat is transferred by the currents. Jet Propulsion Lab

SEASAT SMMR SEA SURFACE TEMPERATURE
SEPTEMBER 23 - OCTOBER 6, 1978

SEASAT SMMR SEA SURFACE TEMPERATURE
JULY 13 - 24, 1978
CHESTER, HUSSEY
NJOKU AND NICHOLS
(1981)
DEGREES CENTIGRADE
4 8 12 16 20 24 28 32

SEASAT SMMR CHANGE IN SEA SURFACE TEMPERATURE
SEPTEMBER MINUS JULY, 1978
CHESTER, HUSSEY
NJOKU AND NICHOLS
(1981)
DEGREES CENTIGRADE
-1.5 -1.0 -.5 0 +.5 +1.0 +1.5

These scientists are preparing a water sampler for launch into the Antarctic Ocean. Woods Hole Oceanographic Institution

The U.S. Coast Guard seized this Cuban trawler because its crew had been accused of taking lobster off the United States continental shelf, fifty miles east of Nantucket Island, Massachusetts. It was released. U.S. Coast Guard

Whales used to be plentiful. Now they need protection or they will become extinct. The Whaling Museum, New Bedford, MA

10

A Storehouse Bursting with Energy

IT'S A CALM DAY at the seashore. The breeze is soft and warm. The surface water is only slightly ruffled. The low waves break into a gently-foaming blanket that spreads smoothly over the sand. There's little indication here that the sea is brimming over with energy. The people who are wading and swimming and sunning themselves aren't aware that there are strong winds blowing several miles offshore. They can't see the swells that are forming. When those swells wash onto the coast, they'll be riding on the back of a high tide. As they crest and break, there'll be no hesitant fingers of water creeping onto the sand with soft hissing sounds. With a crashing, roaring, and pounding the waves will attack again and again. The sights and sounds accompanying that release of energy will be awesome.

If we could capture and harness the energy that's released along the world's shorelines, we'd have a means of supplying electricity to every city, town, and farm on this planet. Finding ways to accomplish this task is one of the most perplexing challenges that man has ever faced. To begin with, most seacoasts don't have the right shape or conditions for the successful operation of wave energy devices. Even where conditions are right, such installations will be difficult to maintain because of the battering of wind and water and the corrosive action of salt, but at the same time the cost of their operation must be kept low, so that people can afford the energy they produce. Finally, anyone building a wave energy plant has to have some idea of the effect it will have on a marine ecosystem that we still don't fully understand.

Presently, there's only one successful working tidal power plant. In northern France, a dam straddles the Rance River estuary. The water that it traps used to rush upriver for fifteen miles. Now it's used to spin the turbine blades that supply electricity to thousands of nearby residents.

Nova Scotia's Bay of Fundy would be another good spot for a tidal power plant, and the Canadian government has studied the advantages and disadvantages of building one there. The project would be expensive and would involve the construction of a complex system of dams and turbines. No one knows the effect the changes in the tidal rhythm would have on the marine plants and animals in the region. On the plus side, it's estimated that the installation would furnish enough power for a city the size of New York.

There is a proposal to harness the steady, dependable force of the Gulf Stream to light and heat homes and businesses all along the southern part of the eastern seaboard. This project would install a series of submerged windmill-like turbines in the Gulf Stream. Protected by a blanket of water, the structures would be little affected by the storms that occur above them. The ecological damage should be very small because the speed of the current would be slowed by only a small amount.

THE MOVEMENT of tides and currents isn't the only potential source of power in the world ocean. Another one lies in the difference in temperature between surface water layers and the bottom layers. The conversion of this temperature differential into electrical energy would yield a continuous supply of power—one much greater than the world could ever use.

OTEC—Ocean Thermal Energy Conversion—is a United States Government pilot project that is studying this energy source. It has operated a sea thermal energy plant in Hawaii since 1979 and is considering another in Guam.

Sea thermal plants have many advantages. Since the ocean itself is their heat storage bin, they need no huge storage areas from which to draw power during times of little or no sunlight. Except for the times when they need maintenance and repair, they can operate twenty-four hours a day. Thermal power doesn't deplete our dwindling coal and oil supplies, and it doesn't burn any air-polluting fuel. It leaves no radioactive waste to cause a disposal problem.

A sea thermal power plant requires a lot of "sophisticated plumbing," but the basic idea is simple. The heat of the surface water is used to turn a low-vapor fluid, such as ammonia, into a gaseous vapor. The vaporized fluid expands and builds up enough pressure to drive a turbine. Cold water is brought up from the sea floor through a pipe. It cools and condenses the fluid, and the cycle starts again.

OTEC engineers are working on the problems that are connected with sea thermal energy plants. Since so much of the maintenance and repair has to be done underwater in hazardous conditions, the structures have to be as durable and dependable as possible. The enormous pipe that's necessary to bring up the cold water might move as much as eight or ten feet as the currents push against it. The shift in location could cause problems on the surface.

Although there are a variety of other technical and environmental problems to solve, OTEC estimates that sea thermal energy plants will be in widespread use by the year 2000. They will be of tremendous help to energy-starved, highly populated areas such as the Hawaiian Islands and Japan.

The movement of water and the sea's temperature differential are two of the most likely ways to obtain power from the world ocean. There have also been many other ideas. Some have been discarded, because they are too expensive or impractical. Others are still being tested. Perhaps a few of them will be put into operation within a few decades. The rest may have to wait for a long time. Engineers have thought of using various forms of baffles, tubes, domes, gates, floats, and man-made whirlpools. A few scientists believe that some energy could be derived from the contact between fresh and salt water at the mouths of big rivers. Others think that a hydrogen based fuel could be extracted from the ocean by using a special sort of bacteria that would be produced through biological engineering.

SOME OF THESE ideas are far-fetched, and it's true that oil will probably be the life blood of the world's power needs for many more years. It's also true, however, that an energy revolution is beginning, and the world ocean is playing a big part in that revolution. To make this source of power even more attractive, some engineers are designing sea thermal systems that will combine energy production with the production of fresh water, of minerals, and of cultured seafood.

Many decades will probably pass before these systems are put into operation. Meanwhile, scientists and engineers are using our present day technology and knowledge to take the first steps away from our dependence upon coal and oil. By tapping the ocean's bursting storehouse of energy, we may eventually be able to stay warm without dirtying our atmosphere. We may be able to operate our factories without depleting our earth's resources. We'll be working with nature, instead of against it.

11

"The Ocean Is Sick, Very Sick"

Roll on, thou deep and dark blue ocean—roll!
Ten thousand fleets sweep over thee in vain;
Man marks the earth with ruin—his control
stops with the shore.

WHEN LORD BYRON wrote those lines in the early 1800s, the ocean appeared to be so vast, so self-healing, so impervious to damage, that man could do it no harm. We now know that the ocean is just as delicate, just as vulnerable, just as subject to Nature's laws as anything else on this planet. Because man has broken some of those laws, the ocean is suffering. In some places, stocks of fish are being depleted. A variety of marine plants are dying out, while others are overproducing. Beaches are eroding as the sea claims them at a steady pace. All around the world, the water in bays and estuaries is being polluted because of man's actions and his neglect.

On March 18, 1967, it became appallingly clear just how quickly an ocean region can be badly damaged. As that day dawned, the Atlantic Ocean off the southern tip of Great Britain sparkled deep blue under the sun. At Land's End, in Cornwall, the hotel owners were preparing to host an influx of tourists from London.

Meanwhile, twenty miles away, the supertanker *Torrey Canyon* was sailing past a group of islands. The visibility was good, the seas calm, and the vessel riding on a high tide. No one anticipated any problems, but a short while later, the ship struck a reef. Thirty-six million gallons of Kuwait crude oil began spilling out of its hold. The damage was so great that there was no chance for any small tankers to rush to the scene to pump out the oil. The entire cargo was lost.

The amount of oil that the supertanker was carrying was only a tiny drop compared to the amount of water that surrounds southern Britain. Nevertheless, it caused extensive damage. Within three days, chocolate-covered slicks had spread over 100 square miles of surface water. Most of the fish, dolphins, and seals, seeming to sense that they were in the presence of an enemy, swam under the slick and escaped. Some were not as fortunate. They were carried toward shore along with a great number of trapped, squawking birds.

The people of Cornwall watched helplessly as the

slick washed onto their golden beaches. The circular action of the waves picked up the sand and mixed it with the oil. Before long, the air was filled with the stench of dead animals. There would be few tourists coming into the area that season. Hotel owners and fishermen alike faced financial ruin.

Three weeks later, the spill arrived at the coast of France. Silently, with no howling winds or crashing waves, the oil-stricken ocean was bringing havoc to the shorelines of two countries. Great Britain tried desperately to limit the damage. Royal Air Force jets dropped napalm bombs on the slick, but crude oil is too thick to burn readily.

"It was," as one reporter stated, "incombustible by anything short of the fires of hell."

The campaign was switched to a fleet of warships armed with chemical detergents. That strategy had mixed results. The oil combined with the detergents to form a mixture that drifted downward, eventually settling on the ocean floor and covering the organisms that normally thrive there. Since such creatures as sea anemones and crabs can't travel to a higher level, they died.

The British finally carried their battle to the beaches. Here the detergents formed a milky liquid that was more toxic to shellfish and tide pool inhabitants than the oil had been. The muck that resulted coated the sand and had to be laboriously removed.

The French attacked the floating oil with a blanket of powdered chalk. It caused part of the slick to sink, but later, when the chalk itself washed onto the beaches, it too had to be removed.

Meanwhile, volunteers were making frantic efforts to save the lives of as many animals as possible. As they smeared the birds with talcum powder to absorb the oil, their hands were bloodied by the clawing of the terrified creatures. Despite an extensive rescue attempt, they saved only a small percent of the stricken birds. The others died, encased in their greasy strait jackets, unable either to float or to fly. Some of them perished when the oil was absorbed by their lungs. Most of them starved to death.

The oil slick was finally dispersed the following June. It had taken no human lives, but it did leave behind it countless dead animals, filthy beaches, great hardships for the people who depended upon fishing or the tourist trade, and a $14 million bill for the repair of the damages.

The *Torrey Canyon* accident also left behind it the frightening knowledge that this was probably only the first of many such incidents. Since then, other supertankers have crashed upon rocks, exploded, broken in half, collided with other ships, and sprung leaks. All of the accidents have had devastating consequences.

Ships aren't the only source of oil pollution. Some of it occurs beause of blowouts and other accidents in offshore drilling areas. Just two years after the *Torrey Canyon* spill, oil began gushing from the sea floor in a drilling site off the coast of Santa Barbara in Southern

California. As it bubbled to the surface and started drifting toward land, workers stood ready to soak up the mess with straw. The cleanup seemed endless as the days passed. Just as soon as the weary volunteers had cleaned off the sand after a high tide, another incoming tide brought more oil with it.

Weeks passed before the crisis was over. A State Fish and Game official said that the spill was the worst disaster ever to hit California wildlife.

The Santa Barbara spill was caused by drilling oil wells into a sea floor near the edge of one of the plates of the earth's crust that has many small fault lines running through it. Although no drilling has been done off Cape Mendocino in Northern California, oceanographers have found that the San Andreas fault there has penetrated into the oil reservoir. The slow seepage can be seen by sonar operators. More and more people are questioning the wisdom of placing drilling rigs in regions that are so geologically unstable.

OIL IS ONLY one of the many substances that are polluting our seas. Some of these substances encourage an excessive growth of a particular organism. These unwanted "blooms" cause biological deserts in which one species of plant or animal crowds out the others in the competition for food and oxygen.

Some blooms seem to occur naturally. One of them involve the algae that sporadically becomes known as a "red tide." One such tide occurred along coastal Florida in November, 1946. As the organism overpopulated the area, it depleted the sea's oxygen supply and caused the death of the many other species that normally live there. The decaying plant and animal matter polluted the water, resulting in still more damage to the ecology. Schools of fish were seen splashing about in a frenzy. Beach-goers coughed and choked as irritating fumes rose from the waves. The water thickened and "ran from the hand in strings like thin syrup," said one observer.

This particular red tide didn't disperse for nine months. It killed much of the sea life and drove away thousands of tourists.

OTHER OCEAN pollutants are absorbed by some animals and plants in the sea but don't seem to harm them. They, however, pass the substance to other species, who are gravely harmed. The pesticide DDT is one of these pollutants. Until its use was restricted, it was widely used to spray crops. Rain later washed it into lakes and rivers, and then into the ocean. Here it was absorbed into the tissues of a few types of fish, then was passed on to the birds that ate those fish. It finally ended up in the predators that fed upon the contaminated birds.

Ocean currents have carried DDT all around the world. It has been found in the seals and penguins of Antarctica. Peregrine falcons, bald eagles, and California brown pelicans carry traces of it. Because of the

DDT, many of these birds are now laying eggs that have very weak shells. They often break before the young are mature enough to hatch. The failure of these creatures to reproduce in large enough numbers has earned them a place on the endangered species list.

THE PRODUCTION of power, whether from nuclear or conventional sources, presents another problem—that of thermal pollution. The water that's heated in all power plants must be released back into the lake or ocean from which it came. The temperature of the water in the immediate area is raised, and the ecology changes. It's not yet known exactly what changes may take place in the surrounding area.

It's also not easy to tell whether or not there really *is* any thermal pollution. A power plant may cause a temperature difference of one or two degrees. In many places along the western coast of the United States, currents can cause temperature variations of five to ten degrees over short periods of time. As warm and cold water sloshes back and forth in a local area, schools of fish move with it according to their needs. These natural changes can easily mask the effect of any man-made changes.

The nuclear age has brought with it another, more dangerous type of pollution—the disposal of nuclear waste. During the 1940s and 1950s, some of these wastes were put into stainless steel drums, which were then encased in cement and dumped into deep ocean trenches. No one ever expected the wastes to leak out. Even if they did, it was thought that they would stay in place because of the lack of water movement at those depths.

Oceanographers now know that there is a lot of water movement in even the deepest parts of the ocean. If the containers start to leak, the wastes will be a hazard for many generations, because the natural decay of some radioactive isotopes takes a long time. During that period, they have a continuing power to pollute.

In 1954, atomic bomb tests in the central Pacific mixed fission products into the upper layers of the ocean. Although the amount of those products was small, the water over a wide area became contaminated. More than ten years later, crabs on nearby islands were still accumulating so much strontium-90 that their meat was not edible. Giant clams had kidneys that were so radioactive they sent the survey geiger counters clear off the scale.

This pollution did have one good side effect. Strontium-90 eventually circulated around the entire Pacific basin. Its presence enabled oceanographers to accurately trace the movements of the mid-ocean gyre. This knowledge is now helping them to gauge the effects of any future ocean pollution.

More knowledge is being gained as scientists try to find better ways to get rid of nuclear waste. One project is centered around the possibility of putting it into weighted cannisters, then burying the cannisters in the sediment that covers the ocean floor. Since nuclear waste continues to give off heat for a long time, some questions have to be answered. Will that heat eventually set up a thermal convection current in the sediment? If so, could that current carry leaking waste back to the surface of the sea floor? Could the current bring the cannister itself to the sediment surface?

To answer all of those questions, scientists must study the way heat travels through the sea floor sediment. By doing so, they will be learning many new facts about the bottom of our world ocean.

COMPARED TO the problems of chemical and nuclear pollution, the problems connected with the dumping of just plain junk doesn't seem very important. Most of the old cars, tin cans, bottles, guns, and worn-out tires just lie on the ocean bottom, creating no hazards. Many smaller fish find shelter from predators by swimming through a broken car window. Some tiny creatures live and breed in soft drink cans.

Some of the junk can present a real danger, however. In 1961, a chemical company was preparing to mine for phosphorus off the coast of southern California. As the members of a crew explored the region, they were startled to find that it was littered with twenty-millimeter shells, many of which were still alive.

The company's officials abandoned the project because they didn't want to endanger anyone's life. Also, as one of them said, "There's enough problems in putting together an undersea mining operation without having to armorplate it."

A similar incident near North Carolina ended in tragedy. A popular fishing spot was once used as a practice artillery range. When the crew of the fishing vessel brought up a live bomb in one of its nets, the jostling caused an explosion. The ship was destroyed and everyone on board was killed.

"Parts of the ocean floor are minefields," wrote an oceanographer, "and there are no charts describing the lethal areas."

THERE ARE few things on earth that are all good or all bad. Even ocean pollution can have beneficial side effects. Carefully measured amounts of sewage released into England's Thames River give needed nourishment to some useful plankton pastures. The striped bass of Washington, D.C.'s Chesapeake Bay also thrive on sewage outfalls. Botanists and biologists are trying to find more ways to put our waste to good use, but again, there are some questions. Will thermal pollution cause toxic blooms or will it turn the region into a tropical paradise teeming with a variety of plant and animal life? How can sewage be handled so it doesn't contaminate shellfish while it's nourishing other types of marine life? How can we discourage destructive blooms while encouraging the growth of nutritious plankton?

The studies continue, and scientists hope that they

will find the right answers before a large-scale calamity occurs. So far, the sea has been largely forgiving of our abuses and neglect. The plastic bottles and other trash that can be seen on the high seas are unsightly, but the water itself is still healthy. It's the water over the continental shelves that's in trouble. We swim and surf in that water. We depend upon it for almost all of our seafood. Since most oceanic life is concentrated on and above the shelves, the things that happen here have a great affect on what happens in the rest of the ocean.

We can't continue to ignore the human deaths that have resulted from the eating of contaminated mussels and clams. There's no escaping the fact that the arsenic and mercury that are dumped into the ocean today will end up on our tables tomorrow. We have to reclaim the estuaries that should be full of life, but where the mud is so polluted all life has disappeared.

The famed ocean explorer Jacques Cousteau has cried out against all the "tossing, shoveling, draining, sinking, leaking, spilling, spurting, and dumping" of polluting substances into our seas.

"The ocean is sick, very sick," he says. "The water we all depend on for a living is becoming a deadly poison. Marine life is getting scarce . . . coral is dying. The marine environment everywhere is getting raped and torn up.

"We must all do our part in stopping the pollution of the sea. If we don't, the ocean will die. And if the ocean dies, mankind will die soon after."

12

The Marriage of Sea and Space

SEA AND SPACE have often been closely connected in the minds of imaginative people. In 1865, Jules Verne sparked an interest in space travel by writing about a fictional journey to the moon. Four years later, he wrote *Twenty Thousand Leagues Under the Sea* and set up waves of speculation about the conditions and creatures that exist in the depths of the world ocean.

In 1969, man finally took that trip into space. In preparation for the venture, the moon had been thoroughly studied and probed by telescopes and other earthbound instruments. It had also been observed by unmanned spacecraft. Neil Armstrong and his fellow astronauts had a good idea of what was facing them at the end of their 214,000-mile flight.

Within ten years, the two Voyager spacecraft had given us a close-up view of Jupiter. They had revealed one of space's most closely kept secrets—the erupting volcanoes on a tiny moon named Io. They had let us hear the eerie, mystical melodies that are played within the Saturnian ring plane.

While man was taking his giant leaps into space, Verne's other visionary adventure had already come true. Auguste Piccard, who in the 1930s had sent balloons into the stratosphere, had gone on to design a submersible called the *Trieste*. It's an elongated vessel that works like a balloon filled with air. Its striped float contains aviation gasoline that provides buoyancy. Below hangs the cabin, which is a seven foot steel sphere.

In 1960, Auguste's son, Jacques, and Naval Lieutenant Don Walsh set a still unbeaten record by using the *Trieste* to descend seven miles into the Pacific Ocean's Mariana Trench. The two men had to have as much courage to make their short journey as the astronauts had to have to take their long one. At that time, only the shallow parts of the continental shelves had been viewed by man. There were no telescopes by which anyone could peer into the ocean's depths. In many ways, the environment of the sea floor was still less familiar to us than the craters of the moon.

The *Trieste* made a perfect landing on an ivory-colored carpet of ooze that had been lain down during the course of thousands of years. Jellyfish, crabs, and worms floated past the windows of the submersible. Within a few moments, a foot-long fish joined them. The sight answered the long-standing question about whether or not a true fish could exist in the deepest parts of the ocean.

Manned submersibles usually are built to descend to depths of two to three miles. To go deeper, they would have to be much more resistant to pressure and have other features that would make them very expensive. At this time, with no danger to man, remote instruments can be used to do whatever jobs are necessary in the deeper parts of the ocean.

Meanwhile, in these submersibles, such as *Alvin*, which is laden with instruments and remote control samplers, scientists are able to climb underwater cliffs as men scale continental mountains. They're able to look around and see the texture of the rocks, ignore the poor samples, and ask the pilot to take only the more valuable specimens. They can view the marine creatures in their natural environment and take extensive pictures and notes. The time spent in a submersible is expensive, so geologists and biologists must make every minute count.

In 1966, *Alvin* played a big part in retrieving a hydrogen bomb that had fallen into the sea off the Spanish coast. Again and again, it carried a team of men into the ocean. The search was exhausting and dangerous.

"It was like flying a helicopter in the Rocky Mountains on a dark night," said one team member. "A mud slide coming down a slope could have given us a very bad time."

Days passed before the men spotted the track that the bomb had made as it slid down a hill. Before they could find the bomb itself, *Alvin*'s batteries almost died, and they had to surface. Later, with the help of CURV, a leashed robot, the bomb was pulled from its precarious perch on the edge of a crevasse.

As IN SPACE, the senses and minds of human beings can play a vital part in the exploration of the sea. Man's walk in space, and his experiments with living and working underwater have proved that humans are the most adaptable of all of earth's creatures. They can cope with garbled speech, nausea, fatigue, disorientation, weightlessness, and pressure and still perform their scheduled tasks and suffer no long-lasting ill effects.

Nevertheless, most of the exploration of space and sea will continue to be done by unmanned or remote instruments and devices. Their "eyes," "ears," and "hands" can work tirelessly, and those that function underwater need none of the time-consuming decompression that prevents pressure-sickness or death in humans. *Gloria* and *Deep Tow* are two of the "seeing ears" that use sonar to explore the structures on the ocean bottom. Robots can build and repair submarine installations and lay pipelines and cables, leaving human beings to do less dangerous work. Someday robots will be used to build offshore airports, loading docks, and power plants. They may even help to construct small communities on the surface of the ocean. Plans for an Aquapolis or a Triton City envision a doughnut-shaped apartment and shopping complex atop a circle of long columns rising from the sea.

Some remote instruments work twenty-four hours a day, all year long. They are the buoys that bob about on the ocean's surface and collect information about weather conditions, surface currents and waves, and the temperature and composition of the water. A newly-designed monster buoy can take one hundred different air and water measurements while continuing to function in winds up to 160 miles per hour and in seas sixty feet high.

The data that's gathered by buoys is often transmitted to the most spectacular of all remote instruments —the earth orbiting oceanographic satellites. The first of these—Seasat-I—was launched in 1978. From its 500-mile high vantage point, Seasat could scan about ninety-five percent of the world ocean every thirty six hours. The Nimbus I is another such satellite. It carries a special instrument that accurately measures the distribution of color on the ocean's surface. This color shows the location of oil slicks, the presence or absence of marine life, the meanderings of currents, and the areas where rivers spread their sediment and pollution into the sea.

Plans are being made for the launching of other oceanographic satellites. TOPEX will measure ocean currents more accurately than has ever been done before.

WITH ROBOTS prowling the ocean and satellites keeping a tireless vigil from space, it may sometimes seem that human beings are being pushed to the sidelines in the exploration of the world ocean. The opposite is true. Only man has the ability to make quick decisions and to direct these remote instruments. Only man has the ability to analyze the data and apply the results to the needs of fishermen, merchants, and sailors. Only man has the curiosity to keep on with his explorations and his probings of the unknown. Most of our discoveries have been made by oceanographers going out on small ships again and again. They have been the ones who have mapped the ocean floor and discovered the mountain ranges, the trenches, the rifts, and the volcanoes.

When a scientist is at sea, he faces the same inconveniences and dangers that any sailor faces on a small vessel. Doing research under such conditions can be very satisfying. It can also be very frustrating. Measurements are hard to take, because the ship is always moving. The sea floor seems to gobble up cameras and other expensive equipment. If the instruments aren't lost, they often become damaged. If there are no replacements onboard, one or more experiments may have to be postponed.

Anything that's placed on a bench or tabletop will slide off unless it's lashed, nailed, screwed, or taped into place. All small items must go into boxes that are secured to the walls. When a scientist is looking at the recordings of an instrument, he may have to wedge his chair into place and hold onto a solid object to keep from being thrown into the knobs and dials and smashing them.

Dredging and coring is done from the fantail of the ship. The men must work only a few feet above water, so the equipment can be raised and lowered with the least chance of damaging it. Getting soaked in the tropics isn't so bad. It's not so pleasant in the Arctic Sea.

Accidents can happen no matter how careful everyone is. Two oceanographers were once leaning against the rail of their research vessel. They were directly on a line between the dredge on the deck and the sheave, or pulley, on the end of a boom through which the dredging wire led to the winch. One of the men grew uneasy. He knew there was little chance of the winch starting up, but nevertheless he took his companion's arm and they moved to another location.

Moments later, for no apparent reason, the winch started to haul in the line. In response, the dredge moved across the deck and tore away the railing against which the men had been leaning a short while earlier.

All oceanographers eventually have some close calls. They all experience the discomforts that occur when a ship is rolling and pitching under their feet. And they all think that their work is well worth it. "I've never lost the wonder and excitement of the sea," said one man.

He went on to tell of his first trip in a submersible. "The whole scene could best be pictured by imagining a series of steep rocky alpines dusted by freshly fallen powder snow and bathed in green moonlight."

Oceanographers aren't working only to satisfy their unending curiosity. They know that their discoveries benefit people. In just one area—that of weather prediction—they may soon be able to save thousands of lives by issuing long-term predictions about floods and droughts.

A few climatologists are currently keeping a close eye on the polar icecaps. They claim that the ice is melting because of a "greenhouse effect" in which heat radiating from the earth is trapped under a thickening layer of carbon dioxide. The melting is causing the sea level to increase slightly. If that increase were to speed up, it could destroy the massive unstable ice sheet in western Antarctica. The oceans could then rise and inundate the coastal regions of the world. Such massive flooding could occur within the next fifty to two hundred years. By monitoring the melting of the icecaps, oceanographers could give plenty of advance warning to people who live in Gulf Coast cities, such as Galveston and New Orleans, plus Savannah, Charleston, New York City, and Boston and other towns and cities all along our Eastern seaboard.

While some oceanographers are keeping an eye out for potential disasters, others are working on improving the lives of human beings. They are constantly seeking new ways to use the resources that the sea holds in such abundance. One enthusiastic researcher said, "Someday we'll be using everything in the ocean except the roar of the surf."

Perhaps considering our current technology, he was exaggerating. Perhaps not. When we finally harness the power of tides and waves, we may someday be using everything in the ocean *including* the roar of the surf.

LYNDON JOHNSON once said, "The depth of the sea is a new environment for man's exploration and development, just as crossing the western plains was a challenge to centuries past. We shall encounter that environment with the same conviction and pioneering spirit that propelled ships from the Old to the New World."

Every oceanographer and space explorer knows that the age of discovery is far from over. In a few years, Voyager II will be sending us pictures of Uranus. At the same time, we may be touching the surface of the abyssal plain as we have touched the surface of the moon.

13

Who Owns the World Ocean?

IT IS the year 2500. The maritime powers of the world have extended their sovereignty to the center of the oceans. Cargo, passenger, military, and research vessels alike must pay tribute as they sail through straits that were once open and free to all, or as they pass from one political zone to another. Long lines of factory ships advance across the sea, suctioning up tons of fish in every mile-long sweep. Beneath them, on the ocean floor, robots are dredging, digging, and vacuuming up the mineral wealth that has lain untouched for millions of years.

Disputes over boundary lines, fishing areas, ocean mining claims, and the rights of passage are keeping the world in a constant state of tension. Toxic chemicals that were used to spray African crops end up killing fish in the Bay of Bengal, and angry words fly between the leaders of the countries involved. As in all cases of pollution, no one wants to accept the responsibility for it or to pay the cost of counteracting it. The widening gap between the "haves" and the "have-nots"—those countries with a large share of the world ocean and those with an insignificant share—is threatening to engage every nation in a global conflict.

There has been a final resolution of one dispute, however. The over-hunting of whales is no longer an issue among the major powers. Whales are now extinct.

The sea used to signify freedom, opportunity, and adventure. It has now become a symbol of greed, pollution, and wasteful exploitation. The battle for its riches may well be the battle that finally destroys our civilization.

THIS SITUATION is not a plot for a science fiction novel. It's a scenario of what may happen if the use of the sea isn't regulated in a way that's fair to everyone. The trends have already been put into motion. We have pollution and a declining animal population within our seas. Peru, Mexico, Ecuador, and Korea are a few of the countries that are eager to extend their jurisdiction over wider and wider strips of ocean territory. The rich, highly industrialized nations are competing for the wealth of the abyssal plains. The time-honored principle of the freedom of the seas is being threatened.

This principle was proclaimed by Hugo Grotius in the seventeenth century. "The sea, like the air," he said, "is not subject to appropriation." Grotius is considered the father of international law, and his proclamation is still largely accepted. The only exceptions are the right of each coastal country to maintain a protective strip of water along its shores, and the fishing agreements to which many countries have agreed.

Wars have been fought in defense of the freedom of the seas. For many years, the infamous Barbary pirates roamed up and down the coast of Africa, plundering, looting, seizing ships. France, Great Britain, and the United States finally defeated the pirates in the early 1800s.

The War of 1812 was fought in part over the issue of freedom of the seas. English ships had been stopping American ships and pressing their crews into the service of their merchant fleet. This practice wasn't mentioned in the peace treaty, but the United States had put up a good fight against the more powerful Royal Navy and the impressment stopped.

THE LAW of the Sea that's evolved over eight thousand years of seafaring still holds fast to the principles of Hugo Grotius. The problem is that, until recently, the ocean was only a hunting ground for fishermen, a highway for ships, and occasionally an arena for battle. Most countries were interested in laying claim to only enough coastal water to prevent their shores from being attacked by enemy ships. A three-mile limit was more than adequate to prevent cannon balls from being lobbed onto their beaches. Within that same three-mile strip there was plenty of marine life to fill the nets of the local fishermen.

The situation has changed dramatically during the last two or three decades. Our technology has far outstripped the traditional laws for the use of the ocean. Our growing need for food, water, minerals, and energy, is forcing us to turn to the sea. We've had to augment the simple idea of freedom of the seas with a complex system of rules and regulations. We need international laws that deal with the responsibility for cleaning up contaminated water; for controlling deep sea mining; for protecting the sea creatures that are being hunted almost to extinction; and for settling disputes that arise over the use of the sea's other resources.

The delegates to the recently-ended United Nations International Conference on the Law of the Sea had to face all of these thorny questions when they first started meeting in 1974. They also had to deal with the ongoing issues of fishing rights and territorial limits. With modern weaponry, the old three-mile limit has long been useless as a protective strip. Would a twelve mile limit be enough? And what about the formation of economic zones, within which each coastal country would have the exclusive right to mine, to drill, and to fish? How wide should such zones be? And does a small island have a right to as wide a zone as a large country? What if the territorial water included the narrow Straits of Gibraltar where the Atlantic meets the Mediterranean or the Bosporus where the Mediterranean meets the Black Sea?

Most nations agreed on the right of "innocent passage" of merchant and passenger ships, but some of them wanted innocent passage to be more sharply defined. Does it include military and scientific vessels on peaceful missions?

Fishing rights are usually bound up with territorial rights, and they can create just as many problems. There have already been cod wars, lobster wars, and tuna wars, some of which have led to armed conflict. Ecuador has seized United States fishing ships that sailed to within 200 miles of its coastline. Countries like Portugal, with slowly developing fishing methods, are concerned about how they can compete with countries that have factory ships.

Some countries are concerned with man's interference with the natural flow of ocean water, because of the possible effects it will have on the ecology of their coastlines. The completion of Egypt's Aswan Dam has almost halted the release of water from the Nile into the eastern Mediterranean. The river used to carry nutrients to the many fish that live near its mouth. It also diluted the very salty water that's found there. The eventual effect of the dam on life all along the eastern shore of the Mediterranean still isn't known.

There is some talk about constructing a sea-level canal across Panama so supertankers can move between the Atlantic and Pacific Oceans. The lock system of the present canal effectively separates the two very different bodies of water. The proposed canal would allow them to flow together. Again, no one can predict what will happen to the intermingling plant and animal life. If there are some bad effects, who will pay the people who depend upon the sea for a living?

IN FORMULATING the new Law of the Sea treaty, the delegates had to contend not only with the differing viewpoints among countries, but also with differing factions within some of those countries. For example, our oil industry would like authority over as much of our continental shelf as possible, while marine scientists push for uniform narrow territorial limits so they can have easy access to foreign shores. Environmentalists want strong restraints on ocean-going transports to cut down on pollution. The shippers who own those vessels balk at any expensive restraints. Our fishermen want to ply their trade with as few regulations as possible, but conservationists want stricter laws to protect endangered species.

The biggest obstacle to the drafting of an acceptable treaty lies in the impending mining of the sea floor.

Several countries, including the United States, have spent a lot of time and money exploring the possible sites for such mining and developing the techniques to do the job. During the conference they had to come face to face with the basic questions of who actually owns the minerals at the bottom of the sea. Should they belong to the people who get to them first? Or should all countries share in the wealth, with most of it going to the countries who are in the greatest need? If so, how much effort will the richer nations be willing to invest in the venture? What about the pollution that may result from the mining? If the underdeveloped Third World nations share in the profits, would they also be required to share in the cost of the cleanup?

THE FINAL draft of the treaty was completed after eight years of discussion, arguments, and compromises. It sets a twelve-mile territorial limit and establishes a 200-mile economic zone for each coastal country. It maintains the right of free passage for merchant ships through important straits. It regulates shipping lanes, environmental protection, scientific exploration, and fisheries.

It also states that all countries shall share in the harvest of the minerals of the ocean floor. This and other clauses pertaining to undersea mining caused the United States to vote against the treaty. Since only three other countries voted the same way, the treaty was passed by an overwhelming majority. When sixty member nations ratify it, we will have a new Law of the Sea.

It's possible that no United States firm will be willing to mine the ocean floor under the terms of this new law. The people who do so must not only be willing to give up a large share of the profits, but also to share their technology and knowledge with any country that wants them.

In spite of the possible problems, the new treaty has gone a long way toward a long-desired goal. If it's ratified, perhaps all nations can finally start working together to use the ocean for the common good.

"We are at one of those rare moments," said one concerned writer, "when mankind can come together to devise means of preventing future conflict and to shape its destiny, rather than to solve a problem that has already occurred or to deal with the aftermath of war. It's a test of vision and of statesmanship."

But will a Law of the Sea really work as long as there isn't any complete agreement on the mining of the sea floor? Former President of the United States Lyndon Johnson said that it must.

"Under no circumstances," he stated, "must we allow the prospects of rich harvests of mineral wealth to create a new form of competition among the maritime nations. We must avoid a race to grab and hold the lands under the high seas. We must ensure that the deep seas and the ocean bottoms are, and remain, the legacy of all human beings."

Will the new Law of the Sea really work? Yes, but

only if we realize that freedom of the sea doesn't mean freedom to pollute, to fish to extinction, and to mine at will. Only if we're willing to open a new chapter in international relations and cooperation. Only when we stop asking, "Who owns the world ocean?" and start asking, "Who will care for the world ocean?"

The answer to the first question is, of course, we all do. More important, the answer to the second question is, we all must.

Bibliography

BORGESE, ELISABETH MANN. *The Drama of the Ocean,* Harry N. Abrams, Inc., New York, 1975

CARSON, RACHEL. *The Sea Around Us.* Oxford University Press, New York, 1961

CORLISS, WILLIAM R. *Mysteries Beneath the Sea,* Thomas Y. Crowell Co., New York, 1970

COUSTEAU, JACQUES. *Challenge of the Sea.* Harry N. Abrams, Inc., New York, 1974

ERICSON, DAVID B. *The Ever Changing Sea.* Alfred Knopf, New York, 1967

IDYLL, C.P., EDITOR. *Exploring the Ocean World.* Thomas Y. Crowell Co., New York, 1969

MARX, WESLEY. *The Frail Ocean.* Ballantine Books, New York, 1967

MENARD, WILLIAM, AND SCHEIBER, JAN, EDITORS. *Oceans, Our Continuing Frontier.* Publishers, Inc., Del Mar, CA, 1976

PICCARD, JACQUES AND DIETZ, ROBERT. *Seven Miles Down,* G. P. Putnam's Sons, New York, 1961

STONE, PETER. *The Great Ocean Business.* Coward, Mc-Cann, & Geohegan, Inc., New York, 1972

THORNDIKE, JOSEPH, EDITOR. *Mysteries of the Deep,* American Heritage Publishing Co., New York, 1980

TROEBST, CORD-CHRISTIAN. *Conquest of the Sea.* Harper and Row, New York, 1962

Index